OPTICAL NETWORKING CRASH COURSE

STEVEN SHEPARD

McGraw-Hill Telecommunications

OPTICAL NETWORKING CRASH COURSE

STEVEN SHEPARD

McGraw-Hill

New York • Chicago • San Francisco • Lisbon
London • Madrid • Mexico City • Milan • New Delhi
San Juan • Seoul • Singapore • Sydney • Toronto

McGraw-Hill

A Division of The McGraw·Hill Companies

1 2 3 4 5 6 7 8 9 0   DOC/DOC   0 9 8 7 6 5 4 3 2 1 0

ISBN 0-07-137208-3

The sponsoring editor for this book was Steven Chapman, the editing supervisor was Michael Brummit, and the production supervisor was Pamela Pelton. It was set in Fairfield by MacAllister Publishing Services, LLC.

R. R. Donnelley & Sons Company was printer and binder.

McGraw-Hill books are available at special quantity discounts to use as premiums and sales promotions, or for use in corporate training programs. For more information, please write to the Director of Special Sales, McGraw-Hill, Two Penn Plaza, New York, NY 10121-2298. Or contact your local bookstore.

 This book is printed on recycled, acid-free paper containing a minimum of 50 percent recycled de-inked fiber.

DEDICATION

For Gary, who helped me see the light and start this great and grand adventure. And for my family, again and always.

CONTENTS

ACKNOWLEDGMENTS

I owe an enormous debt of thanks to the following people for their support during the research and writing of this book: Gary Martin, Barbara Jorge, Christine Troianello, Rich Campbell, Jack Gerrish, Henry Sherwood, Gary Kessler, Joe Cappetta, Mitch Moore, Kirk Shamberger, Peter Southwick, Ken Camp, Mark Fei, Elvia Szymanski, Sue Wetherell, Dave Brown, Dave Hill, Bill Ribaudo, Naresh Lakhanpal, Ali Abouzari, Mary Garilis, Todd Quam, Jorge Perez Cantú, Bob Dean, Phil Cashia, Walt Elser, Jack Garrett, Mike Lawler, Kenn Sato, Cyril Berg, Martha Bradley, Carmine Ciotola, Brian Clouse, Floyd Cross, Mike Diffenderfer, Pathmal Gunawardana, Carol Hrobon, Richie Parlato, Johan Lüthi, Greg Reinhart, Jacob Larsen, Dave Brown, Mary Pascarella, Carla Krebs, and Marta Ramirez.

As always, my family gave me the support and freedom required to make a book like this happen.

I am grateful to my friend and editor, Steve Chapman of McGraw-Hill, for his support of this concept and untiring support in the face of a chronically late manuscript.

Thank you—this book is for all of you.

PREFACE

Optical networking is happening.

In the last two years, optical networking has risen into the public consciousness in many different ways. It has become the next great technological thing — businesses want it, service providers want to sell it, device manufacturers want to provide equipment, and component manufacturers are scrambling to supply pieces and parts to all of them. At the time of this writing, an 18-month backlog on optical fiber and some optical amplifiers exists because of the enormous and unanticipated demand for high-bandwidth optical connectivity. Corporate decision makers are now being placed in positions of having to assess the strategic, tactical, and operational value of optical networking within their own corporations. Given how new the widespread deployment of the technology actually is, they have little to help them pass safely through the technological rapids.

The technical book marketplace is replete with seemingly countless titles, which more than adequately cover the technological details that underlie optical transmission, switching, and networking. Even though this is a relatively new field compared to its copper cousin, it benefits from the expertise of many people who have contributed widely to the broad collection of literature about optical technologies. Far fewer books exist, however, which address the managerial and strategic implications of optical networking. Most books on the subject are targeted at field application engineers, component design engineers, and other highly technical personnel who must deal with the inner workings of optical technology, often at the component level.

This book is not targeted at the same audience because additional books in that space would be redundant. Instead, this book addresses the issues and concerns that face executive decision-makers, project and application managers, chief technology officers, and marketing personnel responsible for the strategic and tactical deployment of properly chosen and optimally designed network infrastructures. It addresses optical networking from a practical point-of-view, making it clear that although optical solutions offer enormous bandwidth and capable solutions, they are not the only answers to evolving transport challenges. The audience that this book is directed toward must make decisions that require careful analysis of technology options. To that end it also describes and compares alternatives such as ISDN, xDSL, cable modems, wireless local loop offerings such as LMDS, MMDS, and satellite and copper-based transport schemes, such as T1, T3, SONET, and SDH. In other words, this book is not another fire hose of technology, but rather a carefully crafted tool to help decision makers with technology choices.

Furthermore, its content is not limited to optical transport. It also covers optical switching, routing, and other related areas of interest.

OVERALL DESIGN

The book comprises four main sections: The Optical Networking Marketplace, Origins and Fundamentals of Optical Networking, Market Players, and Solutions and Applications.

The Optical Networking Marketplace sets the stage for the introduction of optical networking and offers a broad overview of the market, its scope, its functional segments, its position relative to traditional copper-based solutions, the economics of optical technology, the players in the game (high-level), and the applicaiton-related reasons for its success.

Origins and Fundamentals of Optical Networking introduces the underlying technologies—how they work, how they interoperate with traditional, so-called "legacy" technologies, and what lies ahead as they mature and become more commonly deployed than they are today. This section also includes the rather fascinating history of optical signal sources, sinks, the optical fiber itself, and optical switching and routing.

Market Players examines the various segments of the optical networking marketplace and the companies that are populated. Four key segments make up the optical networking marketplace: the users themselves, the service providers, the equipment manufacturers, and the optoelectronic component manufacturers. These four groups populate the optical networking food chain and are equally important in the developing marketplace. Each segment is examined, with detailed description and analysis of each company.

Solutions and Applications is exactly that — a careful analysis of the many ways in which optical networking solves customer problems, creates innovative applications, and offers enhanced competitive advantage to its users.

The book concludes with an analysis of optical networking from the perspective of the customer, with an eye toward its ability to engender value in the relationship between the service provider and the end customer.

Enjoy the book.

Steven Shepard
December 2000, Williston, Vermont

PART ONE

THE OPTICAL NETWORKING MARKETPLACE

A revolution is underway in the telecommunications transport world that will fundamentally change the way network service providers electronically move information from place to place. The revolution is based on a number of factors including new applications with ferocious demands for bandwidth, the migration of those applications from the client's device into the network as *Application Service Providers* (ASPs) evolve, the movement of bandwidth, transport, and switching out of the network core and into the equipment at the edge of the network, a growing need for absolutely survivable transport media, and a blurring of the lines of responsibility that have traditionally defined the players in the network services transport game. This evolution, characterized by the move from copper-based networks to optical fiber, from timeslot-based transport to wavelength-based transport, from traditional circuit-switching to terabit router and all-optical switch-based networks, is redefining the roles of all the players in the network services game and ushering in the era of optical networking.

Powerful forces are afoot driving the growth of optical deployment. The first of these is the unshakeable demand for bandwidth brought about by the growth of broadband systems and high bandwidth applications. According to network consultancy firm Ryan Hankin Kent, Inc. (RHK), communications traffic will grow more than 1700 percent by 2002 over 1998 numbers. In fact, according to Forrester Research, as demand for multimedia services and high-speed Internet climbs, more than 27 million users in the United States will have broadband access by 2003. Consider the impact that this growth will have on the network backbone as the local loop is enhanced with high-speed access technologies such as DSL, cable modems, *Dense Wavelength Division Multiplexing* (DWDM)-enhanced fiber, and broadband wireless, as the additional traffic converges on the network. A little-known, undocumented feature of the backbone will be discovered: It smokes as it struggles with the load.

The second motive force is economic. As bandwidth has become more available, the price for it has plummeted, driving the bits-per-second market into the same category as pork bellies, Louisiana sweet crude, and Costa Rican coffee beans. This commoditization of bandwidth has not gone unnoticed by consumers of it, who predictably want more for less. Say's Law observes that supply creates its own demand in the open market, certainly the case in the optical networking world; demand has grown at incalculable rates as the network has become more and more capable. The greatest challenge facing incumbent network providers is the fact that their existing networks were deployed in an era when they were monopoly providers. As a consequence, the networks were massively over-engineered with little concern for cost. Today, though, as their marketplaces have grown uncomfortably competitive, service providers have come to realize the liability of a legacy infrastructure. Thus, anything that service providers can do to drive their provisioning costs down is welcome; massive optical pipes represent one part of the solution.

A third motive force is the perceived aging of such legacy technologies as *Synchronous Optical Network* (SONET) and *Synchronous Digital Hierarchy* (SDH). Conceived in 1984 and introduced commercially in the late 1980s, these physical layer multiplexing schemes provide a transport standard for services operating at rates higher than DS3 and a carefully designed suite of overhead functions that ensured network-wide management capabilities, survivability, universal and simple payload add-drop, and vendor interoperability in a period when interoperability was more brochure-ware than reality. Over time, however, SONET and SDH have come to be perceived as being somewhat overhead-heavy. Newer technologies move channel monitoring, error detection, and forward error correction down into the DWDM layer, removing the need for a portion of SONET/SDH's rather significant overhead complement. As optical technologies and their accompanying protocols advance, other SONET/SDH capabilities will become redundant, and may fade away.

The fourth motive force, and perhaps the greatest liability of all, is the lack of network management capability in the typical network. SONET or SDH-based networks were innovative technologies when they were created in the early 1980s; today, however, they are considered to be monolithic, difficult to provision, and costly. Furthermore, the service activation aspect of the provisioning process is enormously complex, making rapid response to customer requests for service difficult to accomplish in a reasonable amount of time. Thus, enhanced network management is an area of significant focus for most legacy service providers.

Finally, the profile of the typical application is changing in response to both network and customer evolution, as evidenced by the growth of storage area networks and application service providers. These relatively new players in the network game are poised, by their very nature, to drive massive volumes of traffic into the network core.

TOWARD A NEW NETWORK MODEL

The traditional legacy telecommunications network consists of two main regions that can be uniquely and clearly identified: the network itself, which provides switching, signaling, and transport for traffic generated by customer applications; and the access loop, which provides the connectivity between the customer's applications and the network. In this model, the network is considered to be a relatively intelligent medium, while the customer equipment is usually considered to be relatively stupid.

Not only is the intelligence considered to be concentrated in the network; so too is the bulk of the bandwidth, because traditional customer applications don't require much of it. Between switches, and between offices, however, enormous bandwidth is needed.

Today, this model is changing quickly. Customer equipment has become intelligent, such that many of the functions previously done within the network cloud are now done at the edge. *Private Branch Exchanges* (PBXs), computers, and other devices are now capable of making discriminatory decisions about required service levels, obviating the dependence upon the massive intelligence embedded in the core.

At the same time, the bandwidth is moving from the core toward the customer, as applications evolve to require it. Massive core bandwidth still exists within the cloud, but the margins of the cloud are expanding toward the customer.

The result of this evolution is a redefinition of the regions of the network. Instead of a low-speed, low-intelligence access segment and a high-speed, highly-intelligent core, the intelligence has migrated outward to the margins of the network and the bandwidth, once exclusively a core resource, is now equally distributed at the edge as well. Thus we see something of a core and edge region developing in response to changing customer requirements.

One reason for this steady migration is the well-known fact within sales and marketing circles that products sell best when they are located close to the buying customer. They are also easier to customize for individual customers when they are physically closest to the situation for which the customer is buying them.

EDGE VERSUS CORE: WHAT'S THE DIFFERENCE?

Edge devices typically operate at the frontier of the network, serving as vital service outposts for their users. Their responsibilities typically include traffic concentration, the process of statistically balancing load against available network resources; discrimination, during which the characteristics of various traffic types are determined; policy enforcement, the process of ensuring that required quality of service levels are available; and protocol internetworking in heterogeneous networks. Edge devices are often the origination point for IP services and typically provide less than 20 Gbps of bandwidth across their backplanes.

Core devices, on the other hand, are responsible for the high-speed forwarding of packet flows from network sources to network destinations. These devices respond to directions from the edge and ensure that resources are available across the *wide area network* (WAN) to ensure that quality of service is guaranteed on an end-to-end basis. They tend to be more robust devices than their edge counterparts, and typically have 20 Gbps or more of full-duplex bandwidth across their backplanes. They are non-blocking, and support larger numbers of high-speed interfaces.

As the network has evolved to this edge/core dichotomy, the market has evolved as well. As convergence continues to advance and multiprotocol, multimedia networks become the rule rather than the exception, sales will grow exponentially. By 2003, RHK estimates that the edge switch and router market will exceed $21 billion, while the core market will be nearly $16 billion. In the core, Cisco currently holds the bulk of the market at roughly 50 percent, slightly less at the edge with 31percent. Other major players include Lucent Technologies, Marconi, Nortel Networks, Juniper, Newbridge, Fore, Avici, and a host of smaller players.

It is interesting to note that in order to adequately implement convergence, the network must undergo a form of *divergence* as it is redesigned in response to consumer demands. As we just described, the traditional network concentrates its bandwidth and intelligence in the core. The evolving network has in many ways been inverted, moving the intelligence and traffic-handling responsibilities out to the user, replacing them with the high bandwidth core described earlier. In effect, the network becomes something of a high-tech donut. A typical edge-core network is shown in Figure 1-1.

The core, then, becomes the domain of optical networking at its best, offering massively scalable bandwidth through

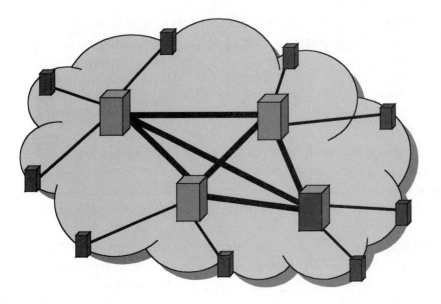

FIGURE 1-1   A typical core edge network

routers capable of handling both high volume traffic and carrying out the QoS dictates of the edge devices that originate the traffic.

The drivers behind this technology schism are similar to those cited earlier. They include:

- The need to create routes on demand between individual users as well as between disparate work groups, in response to the market shying away from dedicated, costly facilities.
- Guaranteed interoperability between disparate protocols.
- Universal, seamless connectivity between far-flung corporate locations.
- Optimum utilization of network bandwidth through the appropriate use of intelligent prioritization and routing techniques.

- Traffic aggregation for wide area transport to ensure efficient use of network bandwidth.
- Granular quality of service control through effective policy and queue management techniques.
- Growing deployment of high-speed access technologies such as DSL, cable modems, wireless local loop and satellite connectivity.

And why is this evolution occurring? Because the closer the services a network provider sells are placed relative to the customer, the more customized, targeted, and immediate those services become. When they are housed in a shared central office, they are much more generalized, catering more to the requirements of the masses and treating the customer as if his or her requirements were commodities. As the network evolves and a clear functional delineation between the edge and the core becomes visible, the role of the central office suddenly changes. In fact, the central office largely disappears. Instead of a massive centralized infrastructure from which all services are delivered (similar to the model employed in legacy data centers), we now see the deployment of hundreds of smaller regional offices placed close to customer concentrations and housing the edge switching and routing technology that deliver the discrete services to each customer on an as-required basis. Those smaller offices are in turn connected to the newly-evolved optical core that replaces the legacy central office and delivers high-speed transport for traffic to and from the customers connected to the edge offices. This is the network model of the future: It is more efficient, and places both the services and the bandwidth that they require where they belong.

Of course, the network management model must now change in response to the new network architecture. Instead of managing centralized network elements—a relatively simple task—the management system must now manage in a distributed environment. This is more complex, but if done properly results in far better customer service because of the immediacy that results from managing at the customer level.

A COROLLARY: THE DATA NETWORK

When data networks first began to be commercially deployed in the 1970s, they followed the same centralized model as the telephone network. Computers, at that time mostly mainframes, were inordinately expensive devices. The applications that they provided were rudimentary at best, and the devices used to access them for data manipulation had the computing power of a water cooler—again, that was a function of the hardware concentrated in the data center.

This model began to change with the introduction of the minicomputer in the late 1970s, and accelerated dramatically with the full-blown arrival of the PC in 1981. Suddenly, the model of computing power concentrated in a central facility (and under the control of a select few techno-druids) was discarded as computing power leaked out of the data center and found its way into the hands of the users themselves. The users soon redefined the model, creating new and innovative uses for the computing power that they now controlled.

It is important to recognize, of course, that the telephone network and the network used by IT personnel to deploy computer services are one and the same. Private networks do exist, but the bulk of all metropolitan and WAN traffic is still carried by telephone company networks. In fact, in the mid-1990s, the telecommunications industry experienced something of an epiphany when they realized the power housed in the global telecommunications network—and began to clamor for its redesign.

From its beginning, the telephone network had been designed largely for the limited transport requirements of voice services. Users of the network required little in the way of bandwidth, so the bulk of the bandwidth, such as it was, was carefully hidden within the largely opaque confines of the network cloud. High capacity T-1, E-1, T-3, E-3 and SONET/SDH trunks moved large volumes of traffic between central offices, but when they were deployed customers were unaware of their existence because it was considered laughable that a customer

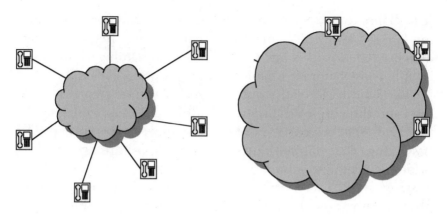

FIGURE 1-2   The migration of bandwidth

could ever require that much bandwidth. Those technologies were designed to be intra-network trunking schemes; customers had no need to know about them.

Of course, times change. It wasn't long before Parkinson's Law[1] kicked in, and soon the edges of the network cloud began to blur as those edges moved outward toward the customer, as shown above. Over time the customer became more informed about the network and its capabilities, the applications began to evolve to take advantage of those capabilities, and soon the demand for ever-more bandwidth was on. The bottleneck began to move as bandwidth seeped out of the transport network's core and began to appear in broadband access technologies such as (initially) ISDN and 56K modems, followed by DSL, broadband wireless technologies, and cable modems. High bandwidth transport was now available not only within the network cloud, but between the cloud and the subscriber as well. And that is precisely where we are today. Because customers *can* transport more traffic into the network cloud, they *will*.

---

[1]Parkinson's Law states that "Work will always grow to occupy the time allotted to it." Its telecommunications corollary states quite accurately that applications will always grow to occupy the amount of bandwidth available to them.

Jean-Baptiste Say, a French economist, is well known for coining Say's Law, which observes that supply creates its own demand. Clearly, this law is at work in the bandwidth markets: As it becomes more available, its consumption goes up.

Bandwidth is not the only network resource that is migrating from the core of the network outward. Let us consider the functions that are carried out in the typical central office. The first things that happen during data transport are aggregation, prioritization, and concentration of traffic. Based on such diverse variables as protocol header information, time of day or physical port number, traffic is segregated by the ingress switch, prioritized, and queued for transport. Once queued, the traffic is transported from one CO to another across high-bandwidth trunks, and multiplexed with other traffic streams to maximize utilization of scarce transport resources. At the destination switch, the traffic is de-multiplexed, queued for delivery, and transmitted from an egress port to the ultimate destination device.

As customer applications have evolved to require more bandwidth in response to its enhanced availability at a reasonable price, as well as dramatic augmentations to the capabilities of *customer-provided equipment* (CPE) such as bridges, routers, switches and multiplexers, the traditional requirement that aggregation, prioritization, and concentration of traffic be done in the network core is no longer valid. Additionally, network protocol models have evolved to the point that traffic prioritization and quality of service guarantees no longer have to derive from central office-based hardware and software. In fact, more and more, modern networks have evolved such that they now comprise two primary layers: an outer edge region, responsible for QoS policy enforcement, and a central core region, responsible for carrying out the mandates of edge devices and transporting traffic at all possible haste across the WAN. One driver behind this evolution is the shift in protocol models as IP-based networks become more and more common, and IP delivers upon its promise of a protocol-agnostic, high-bandwidth, QoS-driven architecture.

THE LOCAL SERVICE PROVIDERS' RESPONSE

The world's local service providers (ILECs, CLECs, PTTs) are watching this network evolution occur, and from their perspective it translates into a number of critical points that they *must* observe and respond to. First of all, the traditional game that they have always played—and to a large degree created the rules for—is changing at a rapid clip. As their customers follow the convergence mantra and orient themselves to focus on services rather than technologies as their primary deliverables, the demands placed on the network are changing in lockstep. No longer do customers necessarily see the service provider network as a source of intelligence and switching capability; instead, many look to it as a high-speed transport mechanism that will guarantee, on an end-to-end basis, the preservation of whatever service characteristics the edge devices attached to the data at the handoff point between the edge and the core. The bandwidth that was for so long concentrated within the fiefdom of the network core has escaped and is now in the hands of the customer, who is using it as a battering ram to attack the core with unfettered traffic flows.

Equally vexing for the service provider is that this evolution runs the potential to relegate them to the role of commodity provider, a role that they are loath to accept. As a result they are scrambling to augment their transport networks with value-added services that will convert them into more of a true *service* provider. One way they are looking to do this is by actively expanding their domain of influence beyond the core to include the edge and beyond. They currently own the bulk of the local loops that are deployed; many of them would like to enter the lucrative CPE game by forming service alliances with equipment manufacturers. This would allow them to offer end-to-end solutions, guaranteeing not only the transport quality but the edge functionality as well. And as optical networking rises to take its place in the bandwidth-rich telecommunications arena, its impact will be profoundly felt as it replaces traditional network technologies.

Service providers and hardware manufacturers alike are also involved in the process of identifying the *service regions* of their network domains. As the products and services they deliver move closer to the customers and the demand for more customized offerings grows, manufacturers and service providers are struggling to define the various network regions within which they operate and the distinct requirements of each of those regions. To date, they have identified five key markets: metropolitan enterprise; metropolitan access; metropolitan backbone; terrestrial long-haul; and submarine long-haul. They are described in detail in the following section.

SERVICE REGIONS OF THE OPTICAL NETWORK

Each of the five network regions—metropolitan enterprise, metropolitan access, metropolitan backbone, terrestrial long-haul, and submarine long-haul—is characterized by specific service requirements that must be met if they are to be effective within each region. Manufacturers and service providers have settled on these divisions of market labor (or some version of them) because together they cover the optical waterfront in terms of products, services, and customer segmentation. Keep in mind that the optical marketplace encompasses both metallic and optical technologies, because traditional technologies are widely deployed and are responsible for feeding the bulk of the traffic into the optical backbone.

Metropolitan Enterprise: The metropolitan enterprise segment covers the range of technologies found within the typical corporate environment. This includes not only the traditional office building complex within which are found large numbers of workers, but campus environments and the network architectures that support telecommuters and the small-office/home-office segment. These would include such technologies as Ethernet, Fast Ethernet, and Gigabit Ethernet; wireless local loop technologies such as *Local Multipoint Distribution System* (LMDS), *Multichannel, Multipoint Distribution System*

(MMDS), *Digital Subscriber Line* (DSL), cable modems, and other high-speed solutions that feed into the optical backbone. Because these technologies "touch" the customer, they tend to be diverse in terms of their bandwidth offerings and service flexibility.

Metropolitan Access: The metropolitan access network region hosts a wide array of services, including service level discrimination, protocol adaptation for purposes of interoperability, traffic aggregation for statistical performance, edge switching, and collection of data for the purposes of billing and administration. At its most fundamental level, the technologies provided here must be relatively service-specific, because they serve as the interface point between the user and the network core. At the very least they must support point-to-point, point-to-multipoint, and virtual networking options, as well as quality-of-service specifics and protocol transparency.

Metropolitan Backbone: The metropolitan backbone network is best described as the network segment that is used for the high-speed, high-volume transport of traffic, usually between corporate facilities, within a metropolitan area. It is typically (but not always) a ring architecture, either a *Unidirectional Path-Switched Ring* (UPSR) for hubbed traffic patterns, a B*i-directional Line-Switched Ring* (BLSR) for distributed traffic patterns, or a mesh arrangement in which rings are interconnected to form a hybrid service solution. The ring usually runs SONET or SDH, and may or may not be further augmented by *Dense Wavelength Division Multiplexing* (DWDM).

Terrestrial Long-Haul: Beyond the world of metropolitan access and transport lies the long-haul interconnect world, which provides long distance transport between cities, countries, and continents. This area is experiencing unprecedented growth: with DWDM, the capacity of fiber is currently doubling every year, and the cost is halving every two years. Both of these are the result of increased channel counts in the DWDM space, the ability to transport as much as 40 Gbps per optical channel, significantly greater optical amplifier bandwidth, better spectrum use efficiency, and lower cost components. They are also the result of demand for diverse long-haul services: wavelength leasing, IP-centric services, *Storage Area Network* (SAN) transport, carrier-to-carrier transport, and metro traffic aggregation and transport. All of these services are failure-sensitive; as a result, route protection and circuit diversity, under the auspices of network management, are critically important.

Submarine Long-Haul: Certainly the most technologically fascinating of the optical applications, submarine brings with it a whole set of challenges that must be dealt with. The greatest of these is geography: because submarine systems typically cross oceans or other large bodies of water, they are logistically difficult to install, maintain, power, and survey. As multinational traffic grows, however, in response to the growth in multinational corporations, submarine optical facilities have become more critical. Interestingly, the companies designing, engineering, and building them such as Global Crossing, Tycom, and Qwest are building them with an eye toward survivability, in many cases creating vast optical rings and meshed network architectures to ensure continued service in the event of a cable failure.

These five market regions represent the end-to-end world of optical networking. Together they satisfy the access and transport requirements of the evolving global optical network. At the access level, the technologies are rich, diverse and growing in bandwidth availability. At the metro transport level, granular bandwidth is the most important characteristic. Finally, across the long haul, high-volume bandwidth and survivability are critical success components.

THE OPTICAL NETWORKING APPLICATIONS SET

Optical networking applications place heavy demands on the underlying physical network. As a result, those networks must be scalable, absolutely reliable, flexible, and must exhibit a high degree of network intelligence. If these capabilities are realized, the network will provide efficient, high-speed provisioning of services, a guarantee of service according to the mandates of whatever service level agreements cover the relationship between provider and consumer of network services, unlimited bandwidth, and financial savings on the cost of operations, administration, maintenance and provisioning functions. Service providers will in turn reap the rewards of their efforts by being able to offer the lowest cost-per-provisioned-bit, better margins and return on investment, increased service revenues, and the ability to provide services to the market faster than their competitors.

Within the five regions of the network described earlier, service providers generally agree that roughly a dozen common applications exist that optical networks lend themselves to. They are the following:

- The long-haul optical packet core
- The regional metropolitan packet core
- The metropolitan multiservice access arena
- Digital video services
- The data center environment
- *Storage Area Networks* (SANs)
- The Ethernet-to-SONET/SDH metropolitan access arena
- Individual metropolitan wavelength services

Each of these will be discussed in the following section.

LONG-HAUL OPTICAL PACKET CORE

The optical packet core is characterized as follows. It must be able to transport massive volumes of packet-based traffic, dependably and reliably, across the wide area fabric of the network. It must have adequate excess bandwidth to respond in real-time to peaks in demand, must be absolutely survivable, and must offer the best possible cost per bit per mile. It must also offer diverse routing capability as a way of dealing with network failures and must offer optical-level management and services capability. This service provides the overall connectivity between metro installations and must therefore be able to adapt to the changing requirements of metro applications, particularly given the fact that those metro applications will evolve according to changing customer demands. In effect, the optical core has the ability to create "virtual fiber"—a form of logical networking and a technique that delivers a more cost-effective solution than multiple *Time Division Multiplexing* (TDM) channels (SONET, SDH or DS-3 technology). By using wavelength-level provisioning, services can be offered at a highly granular level, thus allowing the service provider to live up to its name. The goal of most service providers in the core space is to deliver a seamless end-to-end solution by taking advantage of network intelligence in the optical core, thus making possible the interconnection of metro islands.

Three components emerge in studies of the optical core. The first of these is the transport core itself, which is characterized by long-haul, high-volume, low-cost networking, typically reliant upon DWDM. The second is the optical cross-connect or switching fabric that makes possible the creation of highly-scalable, protocol-and-bit-rate-agnostic transport. The third is a distributed element and network management system.

The products and technologies typically found in the packet core include high-speed, large-scale optical switches and cross-connect devices such as Lucent's LambdaRouter, or Nortel's Optera Connect PX Connection Manager, and the multitude of DWDM transmission products that enrich the optical transport marketplace from such companies as Lucent, Nortel, Ciena, Alcatel, and many others. And as for network management products, most vendors offer element managers that manage the individual components that they sell, but few have stepped up to the challenge of offering a true network-wide management system. This will be discussed in greater detail later.

Regional Metropolitan Packet Core

Because the metro core lies closer to the end-customer, the technologies and capabilities found within it are more complex than the relatively straightforward infrastructure components of the long-haul packet core. It must transport not only the limited requirements of traditional TDM (SONET/SDH) traffic, but must also concern itself with high-speed packet services, video, LAN, and other diverse forms of traffic arriving from and going to customer equipment. It must be able to quickly and easily deploy bandwidth-on-demand to meet the constantly changing needs of a diverse and dynamic customer application base, and provide the seamless interface between the long-haul network and the customer.

Requirements for this region include support for multiple protocol transport, voice-quality reliability for both voice and data services, scalability, and manageability.

The technologies typically found within the metropolitan packet core include integrated, multiprotocol service concentrators that serve as multimedia gateways and provide connectivity for ATM, frame relay, and IP with or without MPLS. They help to reduce the costs of installed hardware and collocation. Other devices include traffic concentration devices that provide high-density concentration of low-speed (usually TDM) services for transport over ATM. The services most commonly found here include packet voice, frame relay, ATM, MPLS-enhanced IP, DSL, VPNs, wireless connectivity, private line, and cable-based connections.

METROPOLITAN MULTISERVICE ACCESS

Similar to the metropolitan packet core, the metro multiservice access arena concerns itself with multimedia, but is more focused on the access side of the networking equation. This region evolved from the realization that traditional SONET or SDH solutions, designed for the limited requirements of legacy traffic, must evolve to become more flexible, faster, and data aware if they are to be cost-effective in the modern multiservice, multiprotocol network. Metro multiservice access is currently the area of greatest change in the modern network.

Four key drivers are causing these changes.

The demand for bandwidth is growing faster than anticipated. IP usage, particularly brought about by the Internet, is at an all-time—and growing—high. Other factors include emerging broadband-rich applications such as video, managed IP services, and carrier access, as well as the growing presence of high-speed LAN technologies, such as Fast and Gigabit Ethernet.

Business application interfaces are getting steadily faster. As applications become more media-rich, the requirement for broadband access becomes crucial. In response, large enterprises are increasing their interfaces to DS3 or E3 and beyond. Smaller businesses are installing T1, E1, DSL, wireless solutions, and cable modems in response to similar demands.

The mix of traffic is becoming more data-heavy. Frame relay, ATM, and IP are becoming the dominant traffic type in most corporate networks because of their speed and versatility, and are in fact growing faster than their voice and dedicated circuit counterparts. At the same time, the conversion to packetized voice is well underway in many companies, and with the arrival of true quality of service protocols on the horizon, its growth will accelerate.

Competitive Local Exchange Carriers (CLECs) *and second carriers need cost-effective data solutions.* SONET and SDH were developed for the legacy, monopoly-minded carriers and are not seen as the ideal solution for high-bandwidth transport by smaller, more nimble (and cost-conscious) carriers. Furthermore, as wavelength division multiplexing technology becomes more cost-effective, it will find its way down to the enterprise level for broadband access, and will work closely with ATM as a solution for the concentration and transport requirements of bursty data traffic.

EXPECTED ACCESS TRENDS

Three important trend areas should be observed by network planners and designers: the evolving network architecture, the changing nature of customer access, and the popularity and positioning of access interface technologies.

Generally speaking, the overall network architecture is not expected to change drastically in the near future. Ring topologies or meshed rings will continue to dominate, in response to growing demands for survivability and hubbed services. Bandwidth will continue to drop in cost, but will be deployed responsibly as the demand arises for it and as competition and migration realities become apparent. Massive fiber build-outs will continue to be the norm, and the fiber will reach closer and closer to the customer as the cost of optoelectronics continues to fall. Finally, the responsibility for aggregation of transported services will move closer to the customer, which will in turn increase the demand for and use of multiservice, integrated access technologies such as ATM.

In parallel with the expansion of the optical cloud, customer access will be characterized by an expanded use of optical technologies such as SONET, SDH, and passive optical networking solutions. These will evolve to include DWDM as time goes on and the technology becomes more widespread and cost-effective. At the same time, smaller businesses will deploy DSL, cable modems, T1, and other broadband solutions.

While we will continue to see growing demand for the latest and greatest access technologies, legacy installations will continue to enjoy success. T-1 and E-1 circuits, for example, are deployed by the hundreds of thousands, and will not be disconnected anytime soon. Therefore, as DSL, cable, wireless, and Fast Ethernet continue to grow, T-1, E-1, OC-n, and STM-n will also be deployed, as well as additional frame relay and ATM solutions.

DIGITAL VIDEO SERVICES

A year ago, this would not exist as a standalone category. However, in the past year video has emerged as a significant traffic component and has become a matter of some concern for service providers whose networks must transport it. Voice and data have traditionally been the principal traffic components found on most networks, but video has emerged in response to demands for rich media from customers. Video requires significant bandwidth and will place inordinately high demands upon existing networks to meet its transport requirements. Most of the major manufacturers have entered the video market, including Cisco, Lucent, and Alcatel. From a customer perspective, a number of issues occur with video services. The most significant of these is cost- the cost of access and transport bandwidth, and the cost of the video CODEC equipment. Equally important is the quality of the delivered service, the ability to integrate new video technology with existing network installations, and the ability to manage the new components of the network and service offerings. Clearly, the closer the optical network lies to the source and destination of the video service, the more cost-effective the ability to deliver those services. Thus, video stands to benefit greatly from the optical access and transport marketplace.

THE DATA CENTER ENVIRONMENT

The data center has evolved from being a relatively invisible, corporate expense center necessary for ongoing operations to a profit center that houses databases and applications critical for strategic positioning, competitive analysis, and rapid response to customer requests for service. As such it has become obvious that a wide spectrum of employees require access to the resources housed in the typical data center, a fact that potentially changes the traffic profile rather significantly. Instead of being a closely-held and highly-specialized resource, the data center becomes a source of information and capability accessed from all over the far-flung corporation. Thus, high-bandwidth access is necessary to support the varied nature of customer accessibility.

THE STORAGE AREA NETWORK (SAN)

Hand in hand with the changing nature of the data center and its perceived value to the corporation is the relatively new concept of the storage area network, or SAN. The SAN houses a corporation's data resources and provides fast, accurate access to them via high-speed technologies such as FibreChannel and high-bandwidth edge routers. Again, optical transport rises to the occasion, providing virtually unlimited bandwidth for the dynamically changing transport requirements of database users and corporate report generators.

ETHERNET TO SONET/SDH METROPOLITAN ACCESS

There was a time when companies were housed in a single, monolithic corporate location—the grand headquarters, as it were. As the workplace has changed, as LAN connectivity has become a routine part of corporate communications, and as the virtual corporation has become a reality, the need to move Ethernet traffic between corporate locations at high-speed has become more necessary. Most major manufacturers now sell Ethernet-to-SONET/SDH conversion devices that bridge the gap between the two. The result is that LAN-like service can be provided between far-flung locations because of the speed of the optical SONET/SDH interconnect. This also preserves and extends the life of two somewhat legacy technologies.

METROPOLITAN WAVELENGTH SERVICES

The magic of being a service provider lies in being able to resell the same resource to many different customers without sacrificing the quality of the delivered service. As optical technology has advanced slowly from the core to the edge of the network to provide high-bandwidth connectivity to the customer found there, it has become obvious that few single customers require the massive bandwidth offered by an entire fiber. Thus, service providers now use such technologies as DWDM to divide the available bandwidth into channels, which they then parcel out to customers as required. This allows them to deploy relatively low-cost, high-bandwidth solutions without stranding an entire fiber with a single customer. At the same time, because these are metropolitan solutions, they are not hindered by the distance limitations that plague long-haul technologies. Thus, relatively low-cost solutions such as *Coarse Wavelength Division Multiplexing* (CWDM) can offer a lower channel count without the cost, challenge, and complexity of amplification and tight channel spacing.

A BRIEF HISTORY OF THE NETWORK

A powerful and largely unanticipated reinvention of the network is underway, a reinvention that is both literally and figuratively turning the traditional telecommunications network inside out. It is the direct result of the repurposing of services and technologies in response to customer demands for solutions and services.

The original telephone network was designed as a fully-meshed network. In other words, if customer A wanted to be able to call customer B, A would call the telephone company and ask for a circuit to be installed between him or herself and B. If customer A now wanted to also be able to call customer C, another circuit had to be installed between A and C. The reader should be able to extrapolate what is happening here and quickly understand the implications of this network design model. The total number of circuits that would have to be installed to provide universal connectivity between (n) customers is calculated with the equation $n(n-1)/2$. Thus, our three customers A, B, and C would require $3(3-1)/2$, or three circuits. A very small community of 16 people, as shown in Figure 1-3, would require 120

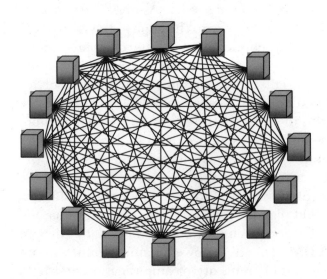

FIGURE 1-3 Full-mesh network

circuits in order for everyone in the town to be able to talk to everyone else in the town. Now imagine a large city of 500,000 people: the number grows to a staggering and incomprehensible 124,999,700,000 (125 billion) circuits. Clearly, this is not a desirable model. Furthermore, this model assumes minimal voice bandwidth requirements. Imagine the impact on the network when broadband transport becomes desirable!

The alternative came in the form of the human telephone operator. The telephone company soon realized that the full-mesh model of networking would rapidly become untenable; in response, the New Haven District Telephone Company came up with the concept of a central exchange in 1878, connecting 21 subscribers to an office where young boys (soon to be replaced permanently by women) established connections per the caller's instructions. This practice continued universally until 1891 when Almon Strowger invented the first mechanical switch to replace the operator. The technology advanced quickly, and by the mid-1970s electromechanical and fully electronic switches had become fairly universal. Note that our network of 16 subscribers, shown in Figure 1-4, now requires only 16 circuits, rather that 120.

These switches, like the operators before them, were housed in central locations for several reasons. First, they were a shared resource and therefore had to be equally accessible to all subscribers within a service area, regardless of whether the service they provided was via human or electromechanical means. Second, they were enormously expensive, and budgets dictated that they should be shared in order to amortize the cost of providing them across a large customer base. As a consequence, the concept of the telephone company central office emerged. Because of the enormous capital required to build the functional center of the network, large, secure buildings were constructed in which to house the hardware that made the local network work properly. All of the intelligence and decision-making capability was housed in the CO; the only network components outside of the CO were the twisted pair local copper loop and the telephones. The telephone had no intelligence; that was a function of the power housed in the CO.

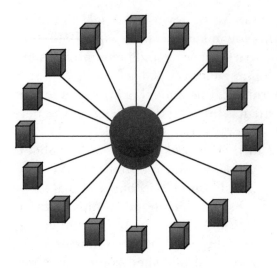

FIGURE 1-4   A centralized network .

THE TRADITIONAL DIGITAL HIERARCHY

The voice network, including both transmission facilities and switching components, was exclusively analog in nature until 1962, when T-carrier emerged as an intra-office trunking scheme. The technology was originally introduced as a short-haul, four-wire facility to serve metropolitan areas. Over the years, it evolved to include coaxial cable facilities, digital microwave systems, fiber, and satellite, and to serve as a long-haul transport solution.

As the network topology improved, so, too, did the switching infrastructure. In 1976, AT&T introduced the 4ESS switch primarily for toll applications, and followed it up with the 5ESS in 1981 for local switching access as well as a variety of remote switching capabilities. Nortel, Siemens, and Ericsson all followed suit with equally capable hardware. In 1983, the first tariff for T-1 was released, and the service was on its way to becoming mainstream.

The T-1 framing and transmission standard is used in North America and Japan, although the remainder of the world uses what is known as the CEPT E-1 standard. CEPT, the *European Council on Post and Telecommunications Administrations*, has largely been replaced today by the *European Telecommunications Standards Institute* (ETSI), but the name is still used occasionally.

E-1 differs from T-1 on several key points. First, it boasts a 2.048 Mbps facility, rather than the 1.544 Mbps facility found in T-1. Second, it utilizes a 32-channel frame rather than 24. Channel one contains framing information and a *four-bit Cyclic Redundancy Check* (CRC-4); Channel 16 contains all signaling information for the frame; and channels one through 15 and 17 through 31 transport user traffic.

A number of similarities between T-1 and E-1 exist as well: channels are all 64 Kbps, and frames are transmitted 8,000 times per second. And whereas T-1 gangs together 24 frames to create an extended superframe, E-1 gangs together 16 frames to create what is known as an ETSI multiframe. The multiframe is subdivided into two sub-multiframes; the CRC-4 in each one is used to check the integrity of the sub-multiframe that preceded it.

One point about T-1 and E-1: because T-1 is a departure from the international E-1 standard, it is incumbent upon the T-1 provider to perform all interconnection conversions between T-1 and E-1 systems. For example, if a call arrives in the United States from a European country, the receiving American carrier must convert the incoming E-1 signal to T-1. If a call originates from Canada and is terminated in Australia, the Canadian originating carrier must convert the call to E-1 before transmitting it to Australia.

When T-1 and E-1 were released in the early 1960s, they offered more bandwidth than any single customer could conceivably use. In fact, they were intended to be used initially as intra-network trunking systems, and therefore represented technology that customers would never see. Of course, that did not last long; soon, customer applications were demanding the levels of bandwidth only possible through the deployment of T-1 and E-1.

In the same way that multiple DS-0 channels are multiplexed to create a T-1 or E-1 in response to demands for

economies of scale, so too can T-1s and E-1s be further multiplexed as a way to share expensive physical facilities. In North
America a digital hierarchy evolved over time to provide standards-based transport for mixed signal components. The North
American Digital Hierarchy is shown here:

SIGNAL LEVEL	BANDWIDTH	# VOICE CHANNELS
DS-0	64 Kbps	1
DS-1	1.544 Mbps	24
DS-2	6.912 Mbps	96
DS-3	44.736 Mbps	672

Similarly, CEPT created its own digital hierarchy for
Europe and the rest of the world, also shown here:

SIGNAL LEVEL	BANDWIDTH	# VOICE CHANNELS
DS-0	64 Kbps	1
E-1	2.048 Mbps	30
E-3	49.152 Mbps	512

The North American and CEPT (now ETSI) digital hierarchies provided a logical multiplexing progression from DS-0
through the highest bandwidth level available within each hierarchy. However, some problems with them did exist. Consider a
situation such as that shown in Figure 1-5. A DS-1 signal is created in the San Francisco central office and is then combined
with three other DS-1s to form an intermediate DS-2. Please
note that in addition to the DS-2s, we must also add overhead
that is used by the network to control and synchronize the payload.

The DS-2 is then combined with six other DS-2s (and more
overhead) to form a composite DS-3 signal. The DS-3 is now
transmitted to the next office.

At the receiving office, the original DS-1 must be dropped
out of the composite signal because it is intended for a receiving device located within that office's purview. In order to locate
the T-1 in question so that it can be dropped, the DS-3 must

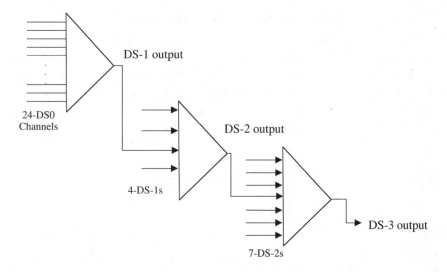

FIGURE 1-5   A DS-1 signal is combined with three others to form a DS-2.

first be fed into a device known as a back-to-back multiplexer (sometimes called a *muldem*) so that it can be disassembled. There, the DS-3 is broken into its composite DS-2s by first backing off the overhead. Next, the proper DS-2 within which the DS-1 is located must be located. The overhead in that particular DS-2 is then removed, the DS-2 is disassembled, and the DS-1 element is dropped out. Another DS-1 may be inserted in its place, after which the entire DS-3 signal must once again be reassembled with appropriate overhead added as required. Needless to say, this process is cumbersome, error-prone, and expensive.

Another problem with the existing hierarchies is that they do not represent transmission speeds beyond a certain point. Practically speaking, the North American Digital Hierarchy, for example, ends at DS-3; anything transmitted faster than that must be done using proprietary multiplexing schemes. This, of course, eliminates vendor interoperability and slows the competitive multiplexing marketplace to a crawl.

Another problem is network management: the interleaved nature of the signal hierarchy makes management of signal elements a complex task.

Clearly, a better system was needed to handle the growing demands of the evolving customer base. That system came with the introduction of the *Synchronous Optical Network* (SONET) and the *Synchronous Digital Hierarchy* (SDH).

THE BIRTH OF OPTICAL NETWORKING: SONET AND SDH

SONET and SDH are internationally recognized multiplexing standards that begin where legacy digital hierarchies (T/E-Carrier) leave off. Proposed initially in 1984, SONET was drafted as a solution to the problem of vendor interoperability following the divestiture of AT&T. Integral to the breakup was the concept of equal access, designed to ensure that customers had the right to choose their long distance carrier from among (at the time) the big three —AT&T, MCI, and Sprint. The problem was that most central offices in the country were replete with Western Electric (a.k.a. AT&T) equipment, which meant that if MCI or Sprint wanted to interconnect with the customers (and therefore equipment) served out of a former AT&T central office, they were obligated to purchase Western Electric hardware to ensure interoperability, because no optical interconnect standards existed at the time. Clearly this did not sit well with MCI and Sprint, who did not want to be obligated to one vendor or another; hence the arrival and rapid success of the SONET standard. Unfortunately, SONET was not applicable to the multiplexing standards in place throughout the rest of the world, and as a result the proposal was rejected by the European standards community in favor of a modified version called the SDH, which matched the European Digital Hierarchy more closely. Over time, an uneasy peace descended between the two, and today they are for the most part fully interoperable. This is important, given the growth in multinational corporations that routinely ship corporate network traffic between continents.

THE ROLE OF SONET AND SDH

Most optical networks today are built upon the fundamental infrastructure introduced in the early 1980s as SONET in North America, Japan, and Korea, or SDH in the rest of the world. SONET and SDH are typically built using a highly-survivable ring architecture that has the ability to heal itself in the event of a fiber cut; today, more than 150,000 SONET and SDH rings are deployed throughout the world.

In SONET, optical fiber serves as the transmission medium, although the majority of the signaling, processing, and switching is accomplished electronically. Thus, for optical signals to be processed in a SONET network, they must pass through the electrical components that process the data, and are therefore subject to a rather complex overhead process that is necessary if the network is to function properly. This overhead process is not only complex, it is also costly and makes the process of scaling the network difficult as growth becomes necessary.

For example, Figure 1-6 shows a high-bandwidth signal originating in the central office shown on the left. The signal is transported to a SONET add-drop multiplexer, a device that enables signal components to be added and dropped at various points along an optical circuit and that performs the *Optical-to-Electrical-to-Optical* (O-E-O) conversion required for long-haul signal delivery across the network. At various intermediate points along the ring, other add-drop multiplexers perform similar functions for the traffic, making it possible to carry multiple data streams from one point to another in the network using time division multiplexing technology that is similar to that used in legacy T- and E-Carrier systems.

One problem with SONET and SDH is that they are relatively bandwidth-limited. When they were first deployed, they offered transport at the almost unimaginable speed of 2.5 Gbps. Today, 10 Gbps systems are routinely deployed, but even they are beginning to appear limited as bandwidth demands continue to climb. Additional bandwidth was required, and it arrived in the form of DWDM, a form of frequency division

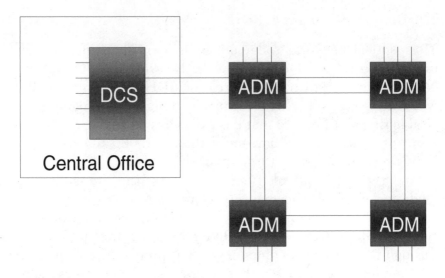

FIGURE 1-6   The O-E-O conversion process

multiplexing designed to increase bandwidth across an optical span that will be discussed later.

SONET TECHNOLOGY OVERVIEW

SONET's lowest transport speed, called *Optical Carrier Level One* (OC-1), is 51.84 Mbps. Higher speeds can be accommodated by allocating even multiples of OC-1 in a series of recognized combinations, as shown:

OPTICAL CARRIER LEVEL	ELECTRICAL LEVEL	BANDWIDTH	SDH STM EQUIVALENT
OC-1	STS-1	51.84 Mbps	—
OC-3	STS-3	155.52 Mbps	STM-1
OC-9	STS-9	466.560 Mbps	STM-3
OC-12	STS-12	622.08 Mbps	STM-4
OC-18	STS-18	933.120 Mbps	STM-6
OC-24	STS-24	1244.16 Mbps	STM-8

OPTICAL CARRIER LEVEL	ELECTRICAL LEVEL	BANDWIDTH	SDH STM EQUIVALENT
OC-36	STS-36	1866.24 Mbps	STM-13
OC-48	STS-48	2488.32 Mbps	STM-16
OC-96	STS-96	4976.64 Mbps	STM-32
OC-192	STS-192	9953.28 Mbps	STM-64
OC-768	STS-768	39813.12 Mbps	STM-256

The SDH standard is similar; although it does not have an equivalent to SONET's OC-1, it does provide an equally simple multiplexing model. In the same way that T-Carrier systems define both a framing standard (T-1) and a multiplexing hierarchy (DS-1), SONET defines both an optical carrier level (OC) and its electrical equivalent, known as the Synchronous Transport Signal (STS). Unlike T-Carrier, the OC/STS/STM levels are exact multiples of the base rate. For example, the bandwidth of a T-1 (1.544 Mbps) is not equal to the bandwidth provided in the 24-64Kbps channels that it contains because of the added overhead required for framing. Similarly, a DS-3's bandwidth is significantly higher than that of 28 DS-1s for the same reasons. In SONET, however, an OC-12 is *exactly* 12 times the bandwidth of an OC-1. This allows SONET (or SDH) to easily and transparently carry payloads at any conceivable speed. A DS-3 fits rather nicely in an OC-1, with a little room left over. A 100 Mbps Fast Ethernet stream can easily be accommodated within a concatenated SONET OC-3, with a little left over. SONET operates under the belief that bandwidth is cheap, and it's a good thing, because it wastes an awful lot of it in unused bytes.

Like T- and E-Carrier, SONET and SDH can be configured as either channelized or unchannelized services. If they are channelized, multiples of the base rate are multiplexed together on a common fiber. For example, an OC-12 requires 622.08 Mbps of bandwidth, but the fastest speed attainable from the service is OC-1. *The user could derive 12 of them, but they would all operate at the fundamental OC-1 rate.* A good analogy is to consider T-Carrier, which works in the same way: the user can get 24-64 Kbps channels, or a single pipe operating at 1.536 Mbps.

If more bandwidth is required for a higher speed application, the channels can be combined as a single, unchannelized circuit. In this case, the aforementioned OC-12 becomes an OC-12c, where the c stands for concatenated, which means chained together. Here, the user is provisioned a single circuit that operates at the aggregate rate of the entire facility.

SONET and SDH are also fully capable of transporting smaller payloads (that is, payloads that require less bandwidth than the base rate provides, such as T-1 or E-1). In this case, the base rate "channel" is broken into smaller payload components called *virtual tributaries* in SONET or *virtual containers* in SDH, each of which are capable of transporting sub-rate services. In SONET, four virtual tributary types exist, shown here with the payload they are each capable of conveniently transporting:

VIRTUAL TRIBUTARY TYPE	PAYLOAD TYPE	BANDWIDTH
VT1.5	DS-1	1.728 Mbps
VT2	E-1	2.304 Mbps
VT3	DS-1C	3.456 Mbps
VT6	DS-2	6.912 Mbps

Similarly, SDH offers four virtual containers, known as *tributary units* (TUs), designed to transport the various payload types found in countries that rely on STM, as shown:

VIRTUAL CONTAINER TYPE	PAYLOAD TYPE	BANDWIDTH
TU-11	DS-1	1.728 Mbps
TU-12	E-1	2.304 Mbps
TU-2	DS-2	6.912 Mbps
TU-3	E-3	49.152 Mbps

OTHER SONET AND SDH ADVANTAGES

Besides providing interoperability (often called "midspan meet," referring to the ability of two multiplexers from different manufacturers to meet in the middle of an optical span and transparently—no pun intended—swap data), SONET and SDH offer a number of other advantages as well. First, they provide a standard multiplexing scheme for services that require bandwidth in excess of the legacy digital hierarchy. Second, they provide embedded, very well designed network management capabilities. Third, they simplify the process of adding and dropping (groom and fill) payload components along the path. Fourth, they facilitate multipoint configurations, allowing payload to be added and dropped at any point in the network with a minimal amount of additional complexity. And finally, SONET and SDH provide for enormous bandwidth, making it possible to transport essentially any payload presented to them.

SONET AND SDH ARCHITECTURES

SONET networks are usually deployed in dual ring configurations as shown in Figure 1-7. In other words, all elements along the span are connected to dual counter-rotating optical rings, one serving as the primary path, the other as a backup. In most cases, the network elements on the ring are add-drop multi-

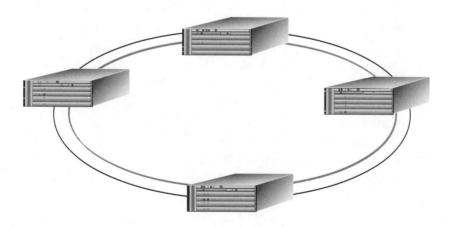

FIGURE 1-7 A sonet dual-ring configuration

plexers, which have the ability to add and drop payload easily at any point along the ring. They are simultaneously connected to both the primary and secondary rings. In the event that the primary ring fails, the devices along the failed path will detect the failure and "wrap," sealing the ring within 50 ms and preventing a catastrophic failure of the circuit. This capability, known as *Automatic Protection Switching* (APS), is one of the key advantages of ring architectures. It is routinely deployed in major networks to eliminate the possibility of a total failure, something that traditional private lines cannot accomplish.

In North America, ring architectures fall into two broad categories. *Unidirectional Path-Switched Rings* (UPSR) are two-fiber architectures. One fiber is designated as the primary path; it is used to transport user traffic under normal operating conditions. For the purposes of this discussion, we will assume that the primary ring operates in a clockwise direction as it moves traffic from multiplexer to multiplexer. The other fiber is designated as a backup span, and other than transporting keep-alive messages, is largely idle under normal operating conditions. It operates in a counter-clockwise direction.

In the event that the primary path should fail due to a fiber cut, the multiplexers on either side of the failure will immediately become aware of the failure through (1) the instantaneous loss of signal and (2) the automatic protection switching overhead, which directs the network to switch to the backup span. This typically occurs within 50 ms, resulting in minimal disruption of the service. However, it does result in another problem. Consider a situation such as the one shown in Figure 1-8, where data is routinely being transmitted from ADM A to ADM B because of the point-to-point fiber between them. When the primary fiber is cut between A and B, transport switches to the backup ring, ensuring an (almost) uninterrupted path. However, the only way that A can transmit to B under this new configuration is by going through E, D, and C on the way to B. This results in an asymmetry in the network that can cause timing and synchronization issues.

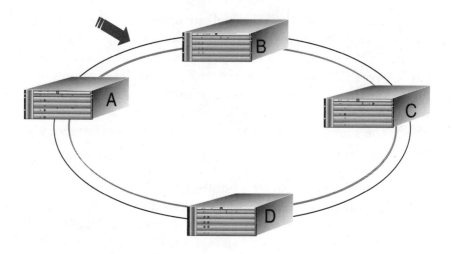

FIGURE 1-8 A sonet UPSR

One solution to this problem is to migrate to what is known as a *Bidirectional Line-Switched Ring* (BLSR). BLSRs are four-fiber transmission architectures in which each pair of fibers serves as a separate bi-directional system. These architectures will be discussed in more detail later in the book.

Beyond SONET and SDH

SONET and SDH are capable physical layer protocols that were designed to handle the legacy time-division multiplexed systems of the mid-1980s. Today, they are beginning to look a bit long in the tooth, largely due to the inordinate percentage of overhead that they are designed around.

The alternative will be optical systems designed to transmit data directly over optical wavelengths. These systems, operating entirely in the optical domain, will be more efficient than SONET/SDH and will provide intelligence at the optical level equivalent to that provided by their TDM ancestors.

This represents the classical inversion of the network described earlier. Most optical networks today rely on frame relay or ATM as their core transport technologies. Routers and switches feed data into the core network, and the core technology establishes the appropriate pathway for data delivery. Evolving networks will be significantly different; instead, IP will serve as the ingress protocol and an intelligent optical core will provide agnostic transport.

One key concern voiced by service providers is the enormous investment in legacy technology that they have. In 1998 alone, incumbent service providers invested nearly $5 billion in SONET hardware, and the number is expected to exceed $20 billion by 2003. Thus new technologies must be backward compatible with legacy installations; failure to be interoperable will result in severe market degradation. On the other hand, emergent greenfield carriers may be in a position to leapfrog legacy technology in favor of next-generation optical solutions, giving them something of a handicap against the larger players.

Hardware providers are taking steps to ensure that next-generation optical networks are as cost-effective as possible for carriers by incorporating new technologies to reduce acquisition costs and minimize the life-cycle cost of advanced-optical systems. According to published reports, optical hardware companies (including Corvis, Ciena, and Nortel's Qtera) claim that their products can reduce the overall cost of ownership by as much as 60 to 80 percent in typical deployment environments.

Transport is only one component of the overall network. Another critical area is switching and routing.

SWITCHING AND ROUTING

Critical components in the overall design of the modern network are switching and routing. They are often confused with one another, and in fact something of a technological *jihad*[1] is

---

[2]Literally, a Holy War based on ideology.

going on between them. They are not the same, nor are their functions. To understand the differences between them (an understanding that is critical to an understanding of the evolving optical networking market), a quick discussion of data communications protocols is in order.

DATA COMMUNICATIONS PROTOCOLS

The functional components of data communications are called *protocols*. Many different forms of them exist. Military protocols define the rules of engagement that modern armies agree to abide by; diplomatic protocols define the manner in which nations will interact and settle their differences; and medical protocols document the manner in which medications are used to treat illness. The word protocol is defined as a set of rules that facilitates communication. Data communications, then, is the set of protocols that governs the exchange of digital data between computing systems.

Data communications networks are often described in terms of their architectures, as are protocols. Protocol architectures are often said to be layered, because they are carefully divided into highly related but non-overlapping functional entities. This division of labor not only makes it easier to understand how data communications works, but also makes the deployment of complex networks far easier. We'll see why in a moment.

Consider this seemingly simple scenario: A PC-based e-mail user in Madrid wants to send a large, confidential message to another user in Singapore. The Singapore user is attached to a mainframe-based corporate e-mail system. In order for the two systems to communicate with each other, a remarkably diverse collection of challenges must first be overcome. Let's examine them a bit more closely.

The first and most apparent challenge that must be overcome is the difference between the actual user interfaces on the two systems. The PC screen presents information to the user in a format that is carefully designed to make it intuitively easy to use.

The mainframe system was designed with the same goal of intuitive ease of use in mind, but because a different company designed the interface for a mainframe host using a different design team, it bears no resemblance to the PC interface. Both are equally capable, but completely different. As a result of these differences, if we were to transmit a screen of information from the PC to the mainframe system, it would be unreadable, simply because the two interfaces do not share common field names or locations.

The next problem that must be addressed is the issue of security. We mentioned earlier that the message that is to be sent from the user in Madrid is confidential, which means that it should probably be encrypted to protect its integrity. And because the message is large, the sender will probably compress it to reduce the time it takes to transmit it.

Another problem has to do with the manner in which the information being transmitted is represented. The PC-based message encodes its characters using a 7-bit character set called ASCII, the American Standards Code for Information Interchange. The mainframe, however, uses an 8-bit code called EBCDIC, the Extended Binary Coded Decimal Interchange Code. What happens when a 7-bit system sends information to a system that only understands 8-bit characters? Clearly, problems will result.

Another problem has to do with the logical relationship between the applications running in the two systems. While the PC most likely supports the e-mail account of a single user, the mainframe undoubtedly hosts hundreds, perhaps thousands of accounts, and must therefore ensure that each user receives his or her mail and *only* his or her mail. Some kind of user-by-user and process-by-process differentiation is required to maintain the integrity of the system.

The next major issue has to do with the network over which the information is to be transmitted from Madrid to Singapore. In the past, information was either transmitted via a dedicated and very expensive point-to-point circuit, or over the relatively slow public switched telephone network, or PSTN. Today, however, most modern networks are packet-based, meaning that messages are broken into small, easily routable pieces, called packets, prior to transmission. Of course, this adds an additional layer of complexity to the process: what happens if one of the packets fails to arrive at its destination? Or, what if the packets arrive at the destination out of order?

Computer networks have a lot in common with modern freeway systems, including the tendency to become congested. Congestion results in delay, which some applications do not tolerate well. What happens if some or all of the packets are badly delayed? What is the impact on the end-to-end quality of the service?

Another vexing problem that often occurs is errors in the bitstream. Any number of factors, including sunspot activity, the presence of electric motors, and fluorescent lights can result in ones being changed to zeroes and zeroes being changed to ones. Obviously, this is an undesirable problem, and there must be a technique available to detect and correct these errors when they occur.

There may also be inherent problems with the physical medium over which the information is transmitted. Different types of media include twisted copper wire pairs, optical fiber, coaxial cable, and wireless systems, to name a few. None of these media are perfect; they all suffer from the vagaries of backhoes, lightning strikes, sunlight, earth movement, squirrels with sharp teeth, kids with BB guns, and other impairments far too numerous to name. When these problems occur, how are they detected? Equally important, how are the transmission impairments that they inevitably cause reported and corrected?

In addition, an agreed-upon set of rules that define exactly how the information is to be physically transmitted over the selected medium must be provided. For example, if the protocol to be used dictates that information will *always* be transmitted on pin 2 of a data cable, then the other end will have a problem, because its received signal will arrive on the same pin that it wants to *transmit* on! Furthermore, there must be agreement on how information is to be physically represented, how and when it is to be transmitted, and how it is to be acknowledged.

Collectively, all of these problems pose what seem to be insurmountable challenges to the transmission of data from a source to a receiver. And although the process is obviously complex, steps have been taken to simplify it by breaking it into logical pieces. Those pieces are protocols. Collections of protocols, carefully selected to form functional groupings, are what make data communications work properly.

Perhaps the best-known grouping of protocols is the *International Organization for Standardization's Open Systems Interconnection Reference Model*, usually called the OSI Model. Composed of seven layers, it provides a logical way to study and understand data communications, and is based on the following simple rules. First, each of the seven layers must perform a clearly-defined set of responsibilities that are unique to that layer. Second, each layer depends upon the services of the layers above and below to do its own job. Third, the layers have no idea how the layers around them do what they do; they simply know that they do it. This is called transparency. Finally, seven is not a magic number. If the industry should decide that we need an eighth layer on the model, or that layer five is redundant, then the model will be changed. The key is functionality.

It is important to understand that the OSI Model is nothing more than a conceptual way of thinking about data communications. It isn't hardware; it isn't software. It simplifies the process of data transmission so that it can be easily understood and manipulated.

The OSI Model relies on a process called enveloping to perform its tasks. If we go back to our earlier e-mail example, we find that each time a layer invokes a particular protocol to perform its tasks, it wraps the user's data in an envelope of overhead information that tells the receiving device about the protocol used. For example, if a layer uses a particular compression technique to reduce the size of a transmitted file, then it is important that the receiving end be made aware of the techniques so that it knows how to decompress the file when it receives it. Needless to say, quite a bit of overhead must be transmitted with each piece of user data. The overhead is needed, however, if the transmission is to work properly. So as the user's data passes down the so-called stack from layer to layer, additional information is added at each step of the way.

Let's now go back to our e-mail example, but this time, we'll describe it within the context of OSI's layered architecture.

The sender's e-mail application hands data down to the uppermost layer of the OSI Model, called the *Application Layer*. The Application Layer provides a set of specific services that have to do with the meaning of the data, such as file transfer, terminal emulation, and data interoperability. This interoperability is what enables our PC user and our mainframe-based user to communicate. The Application Layer converts the application-specific information into a common, canonical form that can be understood by both systems. Examples of canonical forms include X.400 and SMTP for e-mail applications, FTP and FTAM for file transfer, and EDI for data representation.

For our e-mail example, let's assume that the Application Layer converts the PC-specific information to X.400 format and adds a header that will tell the receiving device to look for X.400-formatted content. It then hands the now slightly larger message down to the *Presentation Layer*.

The Presentation Layer offers a more general set of services that have to do with the *form* of the data. These services include code conversion, such as ASCII to EBCDIC translation, compression, and encryption. Note that these services can be used on any form of data: spreadsheets, word processing documents, and rock music can all be compressed and encrypted. Our e-mail message is encrypted, but may also be compressed. It may also have to go through a code conversion as it passes from the PC to the mainframe and back again.

The Presentation Layer now hands the data down to the *Session Layer*, which ensures that a logical relationship is created between the transmitting and receiving applications. It ensures, for example, that our PC user in Madrid receives his or her mail and *only* his or her mail from the mainframe, which is undoubtedly hosting large numbers of other e-mail users.

The Session Layer is also responsible for one form of information integrity. Notice that when you log into your e-mail application, the first thing it does is ask for your login ID, which you enter. It appears in the appropriate field on the screen.

When it asks for your password, however, the password does not appear on the screen. This is because the Session Layer has turned off "local echo" so that your keystrokes do not appear on the screen. As soon as the password has been transmitted, the Session Layer turns echo back on again.

The Session Layer now hands the steadily growing *Protocol Data Unit* (PDU) down to the Transport Layer. This is where we first enter the network.

The Transport Layer's job is to guarantee end-to-end delivery of the transmitted message. It does this by taking into account the nature and robustness of the underlying physical network over which the information is being transmitted. For example, if the network consists of a single, dedicated, point-to-point circuit, then very little could happen to the data during the transmission, because the data would consist of an uninterrupted stream. The Transport Layer would have little to do to guarantee the delivery of the message.

However, what if the architecture of the network is not as robust as a private line circuit? What if this is a packet network, in which case the message is broken into segments that are independently routed through the switch and router fabric of the network? Furthermore, what if no guarantee ensures that all of the packets will take the same route through the network wilderness? In that case, the route actually consists of a *series* of routes between the switches or routers, like several strings of sausage links. In this situation, we have no guarantee that the components of the message will arrive in sequence—in fact, we don't have a guarantee that they will arrive at all. In this case, the Transport Layer has major responsibilities to ensure that all of the message components arrive, and that they carry enough additional information in the form of yet another header—this time on each packet—to allow them to be properly re-sequenced at the destination. The Transport Layer becomes the centerpoint of message integrity. Modern switched networks typically fall into one of two major categories: *circuit switched*, in which the network pre-establishes a path for transport of traffic from a source to a destination, as is done in the traditional telephone network; and *store-and-forward networks*, where the traffic is handed from one switch to the next as it makes its way across the network fabric. When traffic arrives at a switch, it is stored in some form of memory, examined for errors and destination information, and forwarded to the next hop along the path—hence the name. Packet switching is one form of store-and-forward technology.

Once the Transport Layer has taken whatever steps are necessary to prepare the packets for their transmission across the network, they are handed down to the Network Layer.

The Network Layer has two primary responsibilities in the name of network integrity: *routing* and *congestion control*. Routing is the process of intelligently selecting the most appropriate route through the network for the packets; congestion control is the process that ensures that the packets are minimally delayed as they make their way from the source to the destination. We will begin with a discussion of routing.

Modern networks, as shown in Figure 1-9, are often represented as a cloud filled with boxes representing switches or routers. Depending upon such factors as congestion, cost, number of hops between routers and other considerations, the net-

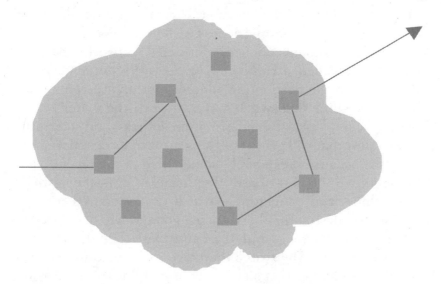

Figure 1-9: The modern network

work will select the optimal end-to-end path for the stream of packets created by the Transport Layer. Depending upon the nature of the network layer protocol that is in use, the network will take one of two actions. It will either establish a single path over which all the packets will travel in sequence, or the network will simply be handed the packets and told to deliver them as it see fit. The first technique, which establishes a seemingly dedicated path, is called *connection-oriented* service; the other technique, which does *not* dedicate a path, is called *connectionless* service. We will discuss each of these in turn.

When Meriwether Lewis and William Clark left St. Louis to travel up the Missouri and Columbia Rivers to the Pacific Ocean, they had no idea how to get where they were going. They traveled with and relied on a massive collection of maps, transcripts of interviews with trappers and Native American guides, and the knowledge of Sacajawea, who accompanied them on their journey. As they made their way across the wilderness of the northwest, they marked trees every few hundred feet by cutting away a large and highly visible swath of bark, a process known as blazing. By blazing their trail, others could easily follow them without the need for maps, trapper lore, or guides.

If you understand this concept, then you also understand the concept of connection-oriented switching. When a device sends packets into a connection-oriented network, the first packet, often called a *call setup* or *discovery packet*, carries embedded in it the destination address that it is searching for. Upon arrival at the first switch in the network, the switch examines the packet, looks at the destination address, and selects an outgoing port that will get the packet closer to its destination. It knows this because presumably, somewhere in the recent past, it has recorded the port of arrival of a packet from the destination machine, and concludes that if a packet arrived on that port from the destination host, then a good way to get closer to the destination is to go out the same port that the arriving packet came in on. The switch then records in its routing tables an entry that dictates that all packets originating from the same source (the source being a virtual circuit address that identifies the logical source of the packets) should be transmitted out the same switch port. This process is then followed by every switch in the path, from the source to the destination. Each switch makes table entries, similar to the blazes left by the Corps of Discovery.

With this technique, the only packet that requires a complete address is the initial one that blazes the trail through the network wilderness. All subsequent packets carry nothing more than a short identifier—a virtual circuit address—that instructs each switch they pass through how to handle them. Thus, all the packets with the same origin will follow the same path through the network. Consequently, they will arrive in order, and will all be delayed the same amount of time as they traverse the network. The service provided by connection-oriented networks is called *virtual circuit service*, because it simulates the service provided by a dedicated service. The technique is called connection-oriented because the switches perceive a relationship, or connection, between all packets that derive from the same source. As with most technologies, a connection-oriented transmission does have a downside. In the event of a network failure or heavy congestion somewhere along the pre-determined path, the circuit is interrupted and will require some form of intervention to correct the problem. Examples of connection-oriented switching include frame relay, X.25 packet-based service, and ATM.

The alternative to connection-oriented switching is *connectionless switching*. In connectionless networks, no predetermined path from the source to the destination exists. We have no call setup packet; all data packets are treated independently, and the switches perceive no relationship between them as they arrive —hence the name. Every packet carries a complete destination address, because it cannot rely on the assistance of a call setup packet.

When a packet arrives at the ingress switch of a connectionless network, the switch examines the packet's destination address, and based on what it knows about the topology of the network, congestion, cost, distance (sometimes called *hop count*), and other factors that affect its routing decisions, it will select an outbound route that optimizes whatever parameters the switch has been instructed to concern itself with. Each switch does the same thing. For example, the first packet of a message, upon arrival at the ingress switch, might be directed out port number seven, because that provides the shortest path to the destination. However, upon closer examination, the switch realizes that although port seven provides the shortest hop count, it is severely congested. Instead, the packet is routed out port 13, which is a longer path but is not congested.

When the second packet arrives, the switch does not realize that it is related to the packet that preceded it. The switch examines the destination address, and then proceeds to route the packet as it did with the preceding one. This time, however, upon examination of the network, it finds that port seven, although clearly the shortest path from the source to the destination, is also no longer congested. It transmits the packet, thus ensuring that packet two will in all likelihood arrive before packet one! Clearly, this poses a problem for message integrity, and illustrates the criticality of the transport layer, which, you will recall, provides end-to-end message integrity by reassembling the message from a collection of out-of-order packets that arrived with varying degrees of delay.

Connectionless service is often called unreliable because it fails to guarantee delay minimums, sequential delivery, or for that matter, *any* kind of delivery! This causes many people to question why network designers would rely on a technology that guarantees so little. The answer lies within the OSI protocol model. Although connectionless networks do not guarantee sequential delivery or limits on delay, they *absolutely* guarantee eventual delivery. Because they are not required to transmit along a fixed path, the switches in the network have the freedom to route around trouble spots by dynamically selecting alternative routes, thus ensuring delivery. If this results in out-of-order delivery, no problem—that's what the transport layer is for. Data communications is a team effort, and requires the efforts of many different layers to ensure the integrity of a message from the transmitter to the receiver. Thus, even an unreliable protocol has distinct advantages. An example of a connectionless protocol is the *Internet Protocol* (IP). It relies on the *Transmission Control Protocol* (TCP), to guarantee end-to-end correctness of the delivered message.

So, how are routing decisions made in a typical network? Whether connectionless or connection-oriented, the routers and switches in the network must take into account a variety of factors to determine the best route for the traffic they manage. These factors fall into a broad category of rule sets called *routing protocols*.

Routing protocols are divided into two main categories—static and dynamic. *Static routing protocols* are those that require a network administrator to establish and maintain them. If a routing table change is required, the network administrator must manually make the change. This ensures absolute security, but is labor intensive and therefore infrequently used. More common are *dynamic routing protocols*, where the network devices themselves make their own decisions about optimum route selection.

Dynamic routing protocols are further divided into two subcategories, centralized and distributed. *Centralized routing protocols* concentrate all of the route decision-making processes in a single node, thus ensuring that all nodes in the network receive the same and most current information possible. Significant downsides to this technique exist: by concentrating all of the decision-making in a single node, the likelihood of a catastrophic failure is dramatically increased. If that node fails, the entire network fails. Second, because all nodes in the network must go to that central device for routing instructions, a significant choke point can result.

Far more common are *distributed routing protocols*. In distributed routing protocols, each device collects information about the topology of the network and makes independent routing decisions based upon what they learn. For example, if a router sees a packet from source X arrive on port 12, it knows that somewhere out port 12 it will find destination X. Thus if a packet arrives on another port looking to be transmitted to X, the router knows that by sending the packet on port 12 it will at least get it closer to its destination. It therefore makes an entry in its routing tables to that effect, so that the next time a packet arrives with the same destination, the switch can quickly consult its table and route the packet properly.

Distributed routing protocols fall into two categories: *distance vector* and *link state*. Distance vector protocols rely on a technique called *table swapping* to exchange information about network topology with each other. This information includes destination/cost pairs that enable each device to select the least cost route from one place to another. On a scheduled basis, routers transmit their entire routing tables on all ports to adjacent devices. Each device then adds new information to its own tables, thus ensuring currency. The only problem with this technique is that it results in a tremendous amount of traffic being sent between network devices, and if the network is relatively static—that is, changes in topology don't happen all that often—then much of the information is unnecessary and can cause serious congestion.

A better solution is the link state protocol. Instead of transmitting entire routing tables on a scheduled basis, link state protocols use a technique called *flooding* to transmit *changes only* to adjacent devices as they occur. This results in less congestion and more efficient use of network resources.

Both distance vector and link state protocols are in widespread use today. The most common distance vector protocols are the *Routing Information Protocol* (RIP), Cisco's *Interior Gateway Routing Protocol* (IGRP), and *Border Gateway Protocol* (BGP). Link state protocols include *Open Shortest Path First* (OSPF), commonly used on the Internet, as well as the *Netware Link Services Protocol* (NLSP), used to route IPX traffic.

Clearly, both connection-oriented and connectionless transport techniques, as well as their related routing protocols, have a place in the modern telecommunications arena. We now turn our attention to the other area of responsibility at the network layer, *congestion control*.

At its most fundamental level, congestion control is a mechanism for reducing the volume of traffic on a particular route through some form of load balancing. This is accomplished in a variety of ways. The simplest technique, used by both frame relay and ATM, is packet discard. In the event that traffic gets too heavy, the switches simply discard excess packets, knowing that the intelligent devices on the ends of the network will detect the loss of information and take steps to have them resent. As drastic as this technique seems, it is not as catastrophic as it appears. Modern networks are heavily dependent on optical fiber and highly capable digital switches; as a result, packet discard, although serious, does not pose a major problem to the network because errors rarely occur.

Other techniques are somewhat more complex than packet discard, but do not result in loss of data. For example, some devices, when informed of congestion within the network, will delay transmission to give the network switches time to process their overload before receiving more. Others will divert traffic to alternate, less-congested routes, or trickle packets into the network, a process known as throttling.

Clearly, the network layer provides a set of centrally important capabilities to the network itself. Through a combination of network protocols, routing protocols, and congestion control protocols, routers and switches provide granular control over the integrity of the network.

Let us return now to our e-mail example. The message has now been divided into packets by the transport layer and delivered in pieces to the network layer, which takes whatever steps are necessary to ensure that they are properly addressed for efficient delivery to the destination. Each packet now has a header attached to it, which contains routing information. The next step, then, is to get the packet correctly to the next link in the network chain—in this case, the next switch or router along the way. The Data Link Layer does this.

The Data Link Layer is responsible for ensuring bit-level integrity of the data being transmitted. When a packet is handed down to the Data Link Layer from the Network Layer, the Data Link Layer wraps the packet in a frame. The frame is made up of several fields that give the network devices the ability to ensure bit-level integrity and proper delivery of the packet from switch to switch.

The beginning and end fields of the frame are called *flags*. These fields, made up of a unique series of bits (0111110), can only occur at the beginning and end of the frame—they are never allowed to occur within the bitstream inside the frame. They are used to signal to a receiving device that a new frame is beginning or ending, which is why their unique bit pattern can never be allowed to occur naturally within the data itself. If the flag pattern *does* occur within the bitstream, it is disrupted by the transmitting device through a process called *bit stuffing*, in which an extra zero is inserted in the middle of the flag pattern. The receiving device has the ability to detect the extra zero and remove it before the data moves up the protocol stack for interpretation. This bit stuffing process guarantees that a false flag will not be interpreted as a final flag and acted upon in error!

The next field in the frame is the *address field*. This address identifies the address of the next switch in the chain to which the frame is directed, and changes at every node. The only address that remains constant is the destination address, safely embedded in the packet itself.

The third field found in many frames is called the *control field*. It contains supervisory information that the network uses to control the integrity of the data link. For example, if a remote device is not responding to a query from a transmitter, the control field can send a "mandatory response required" message that will enable it to determine the nature of the problem at the far end. This field is optional; some protocols do not use it.

The final field we will cover is the *Cyclic Redundancy Check* (CRC) field. The CRC is a mathematical procedure used to test the integrity of the bits within each frame. It does this by treating the zeroes and ones of data as a binary number instead of as a series of characters. It then divides the number by a carefully crafted polynomial value that is designed to *always* yield a remainder following the division process. The value of this remainder is then placed in the CRC field and transmitted as part of the frame to the next switch. The receiving switch performs the same calculation, then compares the two remainders. As long as they are the same, the switch knows that the bits arrived unaltered. If they are different, the received frame is discarded and the transmitting switch is ordered to resend the frame, a process that is repeated until the frame is received correctly. This process can result in transmission delay, because the Data Link Layer will not enable a bad frame to be carried through the network. Thus, the Data Link Layer converts errors into delay.

A number of widely known network technologies are found at the Data Link Layer of the OSI Model. These include the access protocols used in modern local area networks, such as *Carrier Sense, Multiple Access with Collision Detection* (CSMA/CD), used in Ethernet, and token ring. CSMA/CD is a protocol that relies on contention for access to the shared backbone over which all stations transmit, although token ring is more civilized—stations take turns sharing access to the backbone. This is also the domain of frame relay and ATM technologies, which provide high-speed switching.

Frame relay is a high-speed switching technology that has emerged as a good replacement for private line circuits. It offers a wide range of bandwidth and although switched, delivers service quality that is equivalent to that provided by a dedicated facility. It is advantageous for the service provider to sell frame relay because it does not require the establishment of a dedicated circuit, thus making more efficient use of network resources. The only downside of frame relay is that it requires careful engineering to ensure that proper quality of service is delivered to the customer.

Asynchronous Transfer Mode (ATM), has become one of the most important technologies in the service provider pantheon today because it provides the ability to deliver true, dependable, and granular quality of service over a switched network architecture, thus giving service providers the ability to aggregate multiple traffic types on a single network fabric. This means that IP, which is a network layer protocol, can be transported across an ATM backbone, enabling the smooth and service-driven migration to an all-IP network. Eventually, ATM's overhead-heavy quality of service capabilities will be replaced by more elegant solutions, but until that time comes, it still plays a central role in the delivery of service.

Once the CRC has been calculated and the frame is complete, the Data Link Layer hands the frame down to the *Physical Layer*, the lowest layer in the networking food chain. This is the layer responsible for the physical transmission of bits, which it accomplishes in a wide variety of ways. The Physical Layer's job is to ensure bit-level integrity, which includes the proper representation of zeroes and ones, transmission speeds, and physical connector issues. For example, if the network is electrical, then what is the proper range of transmitted voltages required to identify whether the received entity is a zero or a one? Is a one in an optical network represented as the presence of light or the absence of light? Is a one represented in a copper-based system as a positive or as a negative voltage, or both? Also, where is information transmitted and received? For example, if pin two is identified as the transmit lead in a cable, what lead is data received over? All of these physical parameters are designed to ensure that the individual bits are able to maintain their integrity and be recognized by the receiving equipment.

Many transmission standards are found at the Physical Layer, including T1, E1, SONET, SDH, DWDM, and the many flavors of *Digital Subscriber Line* (DSL). T1 and E1 are long-time standards that provide 1.544 and 2.048 Mbps of bandwidth respectively; they have been in existence since the early 1960s and occupy a central position in the typical network. SONET and SDH provide standards-based optical transmission at rates above those provided by the traditional carrier hierarchy. DWDM is a frequency division multiplexing technique that allows multiple wavelengths of light to be transmitted across a single fiber, providing massive bandwidth multiplication across the strand. And DSL extends the useful life of the standard copper wire pair by increasing the bandwidth it is capable of delivering as well as the distance over which that bandwidth can be delivered.

We have now discussed the functions carried out at each layer of the OSI Model. Layers six and seven ensure application integrity, layer five ensures security, and layer four guarantees the integrity of the transmitted message. Layer three ensures network integrity, layer two, data integrity, and layer one, the integrity of the bits themselves. Thus transmission is guaranteed on an end-to-end basis through a series of protocols that are interdependent upon each other and that work closely to ensure integrity at every possible level of the transmission hierarchy.

Of course, OSI is not the only protocol model. In fact, for all its detail and definition, it is rarely used in practice. Instead, it serves as a true model for comparing disparate protocol stacks. In that regard, it is unequalled in its value to data communications.

The most commonly deployed protocol stack is that used in the Internet, the so-called *TCP/IP stack*. Instead of seven layers, TCP/IP consists of four. The bottom layer, called the *Network Interface Layer*, includes the functions performed by OSI's Physical and Data Link Layers. It includes a wide variety of protocols as shown in the illustration.

The *Internet Protocol Layer* (IP) is equivalent in function to the OSI Network Layer. It performs routing and congestion control functions, and as the diagram illustrates, includes RIP, OSPF, and a variety of address conversion protocols.

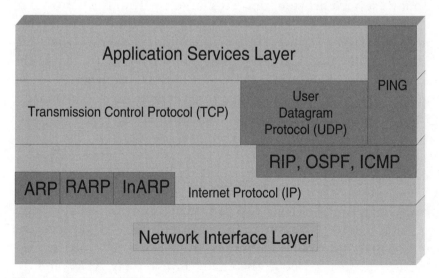

Figure 1-10: The TCP/IP protocol stack

The *Transmission Control Protocol Layer* (TCP) is responsible for message integrity, similar to the service provided by OSI's Transport Layer. It is extremely capable, and has the ability to recover from virtually any network failure imaginable to ensure the integrity of the messages it is designed to protect. For situations where the high degree of protection provided by TCP is considered to be overkill, a corollary protocol, called *User Datagram Protocol* (UDP) is also available at this layer. It provides a connectionless network service and is used in situations where the transported traffic is less critical and where the overhead inherent in TCP poses a problem.

The uppermost layer in the TCP/IP stack is called the *Application Services Layer*. This is where the utility of the stack becomes obvious, because this is where the actual applications are found, such as HTTP, FTP, Telnet, and the other utilities that make the Internet useful to the user.

THE SERVICE PROVIDER'S WORLD: BACK TO SWITCHING AND ROUTING

Today's service provider operates largely within the confines of layers 1-3 of the OSI Model. They provide transmission technologies (Layer 1), switching technologies (Layer 2), and routing technologies (Layer 3). They are striving to reach the upper layers of the protocol model because that is where the customers, and therefore the money, reside. In the meantime, they do an outstanding job of delivering the network infrastructure upon which all services travel.

SWITCHING

Switching, found at layer two, is typically defined as the process of setting up the physical path between network devices and ensuring that the payload entities (frames and cells) arrive properly at the next device in the chain. Frame relay, ATM, and the

PSTN are all examples of switching infrastructures. Switches are not typically considered to be particularly intelligent devices; instead, they often respond to information provided to them by a higher-level protocol layer as they create their pathways through the fabric of the network.

ROUTING

Routing, on the other hand, is found at layer three and is generally considered to be a relatively intelligent process when compared to switching. Routers, for example, have the ability to exchange knowledge about the network with peer routers, thus giving them a global understanding of what is happening across the overall network—something of a Zen-like approach to network traffic control. Because they gain a more in-depth perspective of the state of affairs within the network as a whole, they can make more intelligent decisions about traffic positioning and can thus help to reduce overall congestion within the network.

In the optical world, both switching and routing become extremely important. Optical switching has become quite common today and offers a great deal of promise to network providers that wish to simplify and make their network more efficient. As we will discuss later, true optical switches do not convert an inbound optical signal to electrical in order to process it. Instead, the signal comes into the switch optical, is switched optical, and is then transmitted out the port as an optical signal. This results in tremendous speed and efficiency.

Routing, on the other hand, is equally important, particularly given the ongoing demand for high-speed networking services. As customer demand for bandwidth climbs, more traffic strikes the edge and therefore the core of the network, making high-speed routing a critical element in the optical network. Terabit has become a common term today, and refers to routers that are fast enough to handle the burgeoning requirements of optical networks.

RING ARCHITECTURES IN THE OPTICAL DOMAIN

Today, both SONET and SDH are deployed primarily in ring configurations because of the enhanced survivability that the architectural design provides. Failures come in a variety of forms including optoelectronic failures, digital cross-connect database failures, internal power failures, aerial cable failures, and of course, the infamous backhoe fade. The first three are difficult to protect against, although hardware designers can build in equipment redundancy to reduce the chances of a device-related failure. As it turns out, however, more than 75 percent of all optical network failures result from fiber breakage, either aerial or underground. Thus, anything that can be done architecturally to protect against physical network disruption will go a long way toward ensuring the survival of the network. The most common solution is to deploy redundant paths in the most critical spans of the network, and this is most commonly done using ring topologies.

THE UNIDIRECTIONAL PATH-SWITCHED RING (UPSR)

The unidirectional path-switched ring consists of two fiber rings interconnecting the add-drop devices that make up the network connectivity points. One fiber serves as the active route, while the other, which transmits in the opposite direction, serves as a backup. In the event of a fiber failure between two adjacent nodes, the SONET overhead bytes are used to convey the failure information, as described earlier, one result of which is that the traffic is shunted off to the backup span, usually within 50 ms. The backup span then carries the traffic, this time in the opposite direction, ensuring the integrity of the ring.

Unfortunately, this architecture results in the symmetry problem discussed earlier. Under normal operating conditions, traffic flows from node one to node two to node three and so on. Traffic going from node two to node three only has to traverse a single hop. The other direction, however, is a very different story. Traffic from node three to node two must traverse a four-hop route, resulting in an asymmetric two-way transmission path. For geographically small rings, this is not a problem. For larger rings, however, it can be a significant problem because of synchronization concerns. An application, for example, expecting symmetric transmission and reception, might adjust its timers for equal arrival and departure rates. If it does this, however, it will see data constantly arrive early, while the other end will see it arrive late—or vice-versa. Clearly a better solution might be in order.

BIDIRECTIONAL SWITCHED RINGS

Four-fiber, bidirectional switched rings eliminate the asymmetric problems that characterize unidirectional two-fiber rings. These bidirectional rings can be engineered in two ways: as span-switched rings, or line-switched rings. Both provide high degrees of protection for data traveling on the ring, and both send traffic bidirectionally to ensure symmetrical delay in both transmission directions.

Span-switched rings recover from span failures by switching the bidirectional traffic on the active pair of fibers to the backup pair, but only between the nodes where the failure occurred. This is clearly a better overall design than the two-fiber UPSR, although it does have one common failing: all too often the active and backup fiber pairs are found in the same conduit, which means that a truly diligent backhoe driver WILL take down the ring by cutting all four fibers.

An alternative that offers better survivability is the *Bidirectional Line-Switched Ring* (BLSR). If all four fibers are cut between two adjacent nodes, both the active and the protect path signals wrap to preserve the integrity of the ring. Thus, the BLSR is a more popular configuration in the SONET world, although UPSRs are still more common outside of North America. In fact, four-fiber BLSRs are often seen deployed for long-haul applications, and are usually diversely routed as additional protection against physical failure.

CHARACTERISTIC	UPSR	2F BLSR	4F BLSR
Delay	Asymmetric	Symmetric	Symmetric
Deployment	Metro	Metro	Long-Haul
Multiple Failure Protection	No	No	Yes
Initial Cost	Medium	Medium	High
Augmentation Cost	Low	Medium	Low
Complexity	Low	High	Medium
Efficiency	Medium	Medium	High

AN ALTERNATIVE: THE TWO-FIBER BLSR

Within metropolitan areas, the two-fiber BLSR is becoming a common deployment alternative. It guarantees survivability by gerrymandering available bandwidth. Signals are usually transmitted on both fibers simultaneously. Each multiplexer on the ring is configured to seek its primary signal from one ring or the other, and each span between multiplexers carries configured capacity for both active and protect traffic, to ensure bandwidth availability in the event of a catastrophic failure of the ring.

AMPLIFICATION AND REGENERATION

Over distance, electrical signals tend to weaken and become noisy when transmitted over metallic facilities. The weakness is an issue, because it limits the distance a signal can be transmitted; the noise is a problem because it can mask the integrity of the original signal.

To overcome the problem of weak signals, the signals are often amplified every few thousand feet. This technique does have a downside, however, because amplifiers are signal agnostic—they amplify not only the data signal, but the noise as well. Clearly, this is undesirable; amplifiers are incapable of distinguishing between signal and noise.

To eliminate this problem in digital systems, signals are regenerated. Because digital signals consist of combinations of zeroes and ones represented by a discrete number of easily recognizable pulses, a regenerator has the ability to relatively easily reconstruct the original zeroes and ones, square them up nicely, and send them on their way. Thus digital systems have the ability to absolutely and with great integrity recover the original signal and leave the noise behind.

In optical systems, the same problems occur, although distances are significantly greater and electrical noise is not an issue (although other impairments are). In SONET and SDH systems, signals are regenerated by add-drop multiplexers and other network devices that receive the optical signal, convert it to an electrical signal, regenerate, convert it *back* to an optical signal, and transmit it back onto the fiber ring. Needless to say, this is a complex and expensive task. In modern systems, the optical-to-electrical-to-optical regeneration process has been replaced by all-optical amplification, which is less complex, dramatically faster, and in the long run significantly less expensive. This technique will be discussed later in the book.

BANDWIDTH MULTIPLICATION IN OPTICAL SYSTEMS

In SONET and SDH systems, bandwidth can be increased through a relatively expensive upgrade process that involves the retrofitting of hardware throughout the ring. A recent alternative that has emerged is DWDM . This is not a new technology, by any means. It is simply a new application of frequency division multiplexing, which is a rather old technology! In the case of DWDM, however, the optical signal is subdivided on a wavelength basis, and individual wavelengths, or lambdas, are assigned to different customers, thus allowing them to share the same optical span. This technology will be discussed in detail later in the book, but suffice it to say that DWDM offers remarkable economies of scale to service providers, because it enables them to more efficiently manage and deploy their network resources.

SUMMARY

Clearly a demand for the inordinately high bandwidth that optical networking brings to the pantheon of capabilities offered by the traditional service provider exists. Legacy systems such as T1/E1, SONET, and SDH have long been the mainstays of corporate data transport, but their relatively limited bandwidth and inflexible nature have now relegated them to the legacy technology world. Optical networking, with its high bandwidth, survivability, cost effectiveness and flexibility stands poised to redefine data transport. In the sections that follow we will examine the process by which this will occur.

PART TWO

FROM COPPER TO GLASS

OVERVIEW OF OPTICAL TECHNOLOGY

In 1878, two years after perfecting his speaking telegraph (which became the telephone), Alexander Graham Bell created a device that transmitted the human voice through the air for distances up to 200 meters. The device, which he called the Photophone, (see Figure 2-1) used carefully angled mirrors to reflect sunlight onto a diaphragm that was attached to a mouthpiece. At the receiving end, the light was concentrated by a parabolic mirror onto a selenium resistor, which was connected to a battery and speaker. The diaphragm vibrated when struck by the human voice, which in turn caused the intensity of the light striking the resistor to vary. The selenium resistor, in turn, caused the current flow to vary in concert with the varying sunlight, which caused the received sound to come out of the speaker with remarkable fidelity. This represented the birth of optical transmission!

There were earlier demonstrations of optical transmissions. In 1841, Swiss Physicist Daniel Colladon, and later, in 1870, physicist John Tyndall, well-known for his work on the properties of gases, demonstrated that a beam of light would follow (for the most part) a stream of water issuing from a container, showing that the air-water interface would reflect most of the light back into the stream. Ten years after Tyndall, William

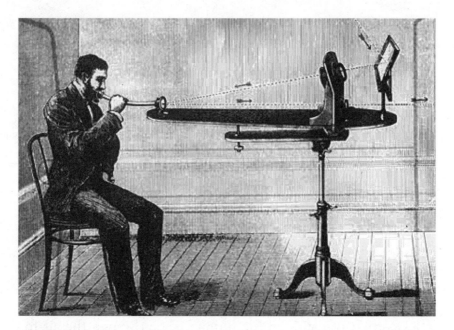

FIGURE 2-1   The Bell Photophone

Wheeling of Concord, Massachusetts (who later became a well-known hydraulics engineer, of all things), created a practical application for Tyndall's demonstration when he used highly reflective metal pipes (see Figure 2-2) to carry the brilliant light from a carbon arc lamp into various rooms in a house. His technique never proved to be commercially practical, but it was the first attempt to pump light for practical reasons, and was the first demonstration of an actual lightguide—a concept that would later be perfected with the development of fiber optics.

It is clear that our fascination with pumped light stems from a point deep in the annals of history, but it has only been relatively recently that optical science has been perfected to the point that optical transmission and its corollary optical switching technology have become not only marketable but capable of redefining the nature of data networking.

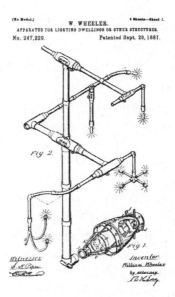

FIGURE 2-2   Tyndall's light pipes

TOTAL INTERNAL REFLECTION

In each of their experiments, Colladon, Wheeler, and Tyndall relied on a phenomenon called *total internal reflection,* which is fundamental to an understanding of how optical transmission works. It is therefore a painful reality that we must delve into optical physics for a moment. This will only be slightly uncomfortable, so go with me on this.

Everyone at one time or another has seen the image of a stick appearing to bend when it is inserted in water. This phenomenon, called *refraction,* occurs because of a difference in the *refractive index* between the air and the water. The refractive index is a measure of the ratio between the speed of light in a vacuum (actually, today measured through the air) and the speed of light in the other medium. Light travels slower in physical media than it does when transmitted through the air, so given that the refractive index [n] is measured as the following:

$$\frac{\text{Speed of light[c] in a vacuum}}{\text{Speed of light [c] in another medium}}$$

The refractive index for any other medium will always be *greater* than one.

So why do we care about this? Because the light actually bends when it encounters the interface between media with different refractive indices. So for example, when a light source shines light into a glass fiber, the light bends as it passes from the air into the glass. The degree that it bends is a function of two things: the difference in refractive index between the two media, and the angle at which the light strikes the glass, known as the *angle of incidence* (see Figure 2-3). This angle is measured from the centerline of the medium, a line that runs perpendicular to the entry surface.

The relationship between the angle of incidence and the angle of refraction is called *Snell's Law*. It becomes important in fiber systems because of the criticality of having the light enter the fiber from the source at as narrow an angle of incidence as possible. If the angle of incidence is too high, the light can actually escape from the glass, resulting in severe signal loss. According to Snell's Law, if the angle of incidence is too high, then refraction will not take place. Put another way, if the light strikes the interface between air and glass (passing into a material with a higher refractive index) at a steep enough angle, the light will not escape but will instead be reflected back into the glass. This process, shown in Figure 2-4, is called *total internal reflection*. It is the basis for transmission through optical fiber. The more of the light that can be kept inside the fiber, the better the integrity of the transmitted signal.

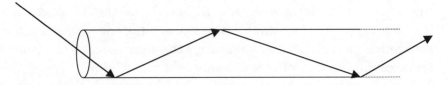

FIGURE 2-3   Angle of incidence

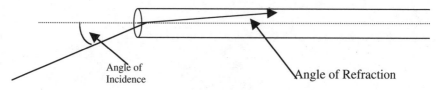

Angle of
Incidence

Angle of Refraction

FIGURE 2-4   Author, need figure caption

Consequently, the angle of the incident light impinging upon the face of the fiber, often called the acceptance angle or numerical aperture, is rather important if the signal is to be transmitted over any significant distance. Therefore, laser assemblies must be carefully crafted to ensure that the face of the laser that will generate the signal to be transmitted is aligned as closely as possible with the face of the fiber—particularly the core of the fiber where the light actually travels. When we consider that modern single-mode fiber has a core diameter of approximately eight, then we realize that the laser itself must be approximately that diameter if it is to direct the bulk of its light output down the core of the fiber. This takes on meaning when we note that a human hair has a diameter of approximately 50 microns! Even in the best systems, about four percent of the signal is lost at these air/glass interfaces. This is known as *Fresnell Loss.*

Because of its design, optical fiber serves as a nearly perfect light guide. It consists of two layers: an inner core through which the bulk of the light travels, and an outer cladding that serves to keep the light in the core (see Figure 2-5).

In typical optical fiber, the refractive index of the core is slightly—but only slightly—higher than the refractive index of the cladding. Thus, the two refractive indices, working closely with the angle of incidence of the transmitted light, ensure that minimal light escapes from the core. Were it not for the cladding, much of the light would escape from the core and be lost over distance.

All right, enough physics for a moment—back to the industry.

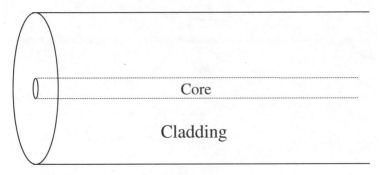

FIGURE 2-5   Fiber structure

LATER DEVELOPMENTS IN OPTICAL TRANSMISSION

Optical transmission of light saw its first practical application with the creation of the fiberscope, simultaneously invented in the 1950s by Narinder Kapany at the Imperial College of Science and Technology in London, and Brian O'Brian of the American Optical Company. The device, designed to conduct an optical image from a source to a destination, relied on glass fiber and was used for industrial inspections and medical applications. Luckily, the distances traversed were quite short, because the unclad fibers of the fiberscope suffered tremendous loss, even over such short distances.

The next major advancement in optical transmission came with the creation of high-quality, low-cost light sources. These came initially in two forms: as *light-emitting diodes* (LEDs) and as *laser diodes* (LDs). Gordon Gould, a graduate student at Columbia University, did the initial work on coherent laser light in the 1950s. Researchers at Bell Laboratories in Murray Hill, New Jersey took it to the next level by attempting to craft practical applications for the technology. In 1962, the first semiconductor lasers were created; these became the centerpoint for transmission over fiber optics.

Of course, optical transmission in the early days was limited in its capabilities. Even though modulated light has information carrying capacity that is orders of magnitude greater than radio, its earliest incarnations suffered tremendous signal loss over distance because of impurities in the glass and the limitations of the optoelectronics that drove the light signal.

Consider the following analogy. If you look through a two-foot square pane of window glass, it is absolutely clear—if the glass is clean, it is virtually invisible. However, if you turn the pane on edge and look through it that way, the glass is dark green. Very little light passes from one edge to the other. In this example, you are looking through two feet of glass. Imagine trying to pass a high-bandwidth signal through 40 or more kilometers of that glass!

In 1966, Charles Kao (see Figure 2-6) and Charles Hockham at the UK's Standard Telecommunication Laboratory (now part of Nortel Networks) published seminal work, demonstrating that optical fiber could be used to carry information provided that its end-to-end signal loss could be kept below 20

FIGURE 2-6   Charles Kao in his laboratory in 1966.

dB per kilometer. Keeping in mind that the decibel scale is logarithmic, 20 dB means that 99 percent of the light would be lost over each kilometer of distance. Only one percent would actually reach the receiver—and that's a one-kilometer run! Kao and Hockham proved that metallic impurities in the glass such as chromium, vanadium, iron, and copper were the primary reasons for such high levels of loss. In response, glass manufacturers rose to the challenge and began to research the creation of ultra-pure products.

In 1970, Peter Schultz, Robert Maurer, and Donald Keck of Corning Glass Works (now Corning Corporation) announced the development of a glass fiber that offered better attenuation than the recognized 20 dB threshold. Today, fiber manufacturers offer fiber so incredibly pure that 10 percent of the light arrives at a receiver placed 50 kilometers away. Put another way, a fiber with 0.2 dB of measured loss delivers more than 60 percent of the transmitted light over a distance of 10 kilometers. Remember the windowpane example? Imagine glass so pure that you could see clearly through a window 10 kilometers thick. Truly incredible.

FUNDAMENTALS OF OPTICAL NETWORKING

At their most basic level, optical networks require three fundamental components: a source of light, a medium over which to transport it, and a receiver for the light (see Figure 2-7). Additionally, there may be regenerators, optical amplifiers, and other pieces of equipment in the circuit. We will examine each of these generic components in turn.

OPTICAL SOURCES

Today the most common sources of light for optical systems are either light-emitting diodes or laser diodes. Both are commonly used, although laser diodes have become more common for

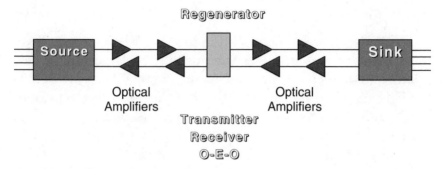

FIGURE 2-7 A generic optical network

high-speed data applications because of their coherent signal. While lasers have gone through several iterations over the years including ruby rod and helium-neon, semiconductor lasers became the norm shortly after their introduction in the early 1960s because of their low cost and high stability.

LIGHT-EMITTING DIODES (LEDs)

Light-emitting diodes come in two varieties: *surface-emitting LEDs* and *edge-emitting LEDs*. Surface emitting LEDs give off light at a wide angle, and therefore do not lend themselves to the more coherent requirements of optical data systems because of the difficulty involved in focusing their emitted light into the core of the receiving fiber. Instead, they are often used as indicators and signaling devices. They are, however, quite inexpensive, and are therefore commonly found.

An alternative to the surface-emitting LED is the edge-emitting device. Edge emitters produce light at significantly narrower angles and have a smaller emitting area, which means that more of their emitted light can be focused into the core. They are typically faster devices than surface emitters, but do have a downside: they are temperature-sensitive, and must therefore be installed in environmentally-controlled devices to ensure the stability of the transmitted signal.

LASER DIODES

Laser diodes represent the alternative to LEDs. A laser diode has a very small emitting surface, usually no larger than a few microns in diameter, which means that a great deal of the emitted light can be directed into the fiber. Because they represent a coherent source, the emission angle of a laser diode is extremely narrow. It is the fastest of the three devices.

Many different types of laser diodes exist. The most common are the *electro-absorptive modulated laser* (EML), which combines a continuous wave laser with a modulating shutter device; the distributed feedback laser, which has an integrated grating assembly to maintain a constant output frequency; and a *vertical cavity surface-emitting laser* (VCSEL, pronounced *vick*-sel), which produces light from a round spot, resulting in a beam of light that is less prone to spread than a typical surface-emitting laser's output. VCSELs are low-power, low-cost, multifrequency devices. Finally, Fabry-Perot lasers are older devices that suffer a number of problems and are less commonly used. They tend to emit light at multiple, closely-spaced wavelengths and are commonly called multimode lasers.

The emission characteristics of all three devices are shown schematically in Figure 2-8. The surface-emitting LED has the widest emission pattern, followed by the edge emitter. The laser diode represents the most coherent and therefore effective light generator. In fact, the graph of the output signal of an LED versus that of a laser is rather dramatic, as shown in Figure 2-9.

OPTICAL FIBER

When Peter Schultz, Donald Keck, and Robert Maurer began their work at Corning to create a low-loss optical fiber, they did so using a newly-crafted process called *inside vapor deposition* (IVD). Whereas most glass is manufactured by melting and reshaping silica, IVD deposits various combinations of carefully selected compounds on the inside surface of a silica tube. The tube becomes the cladding of the fiber; the vapor-deposited

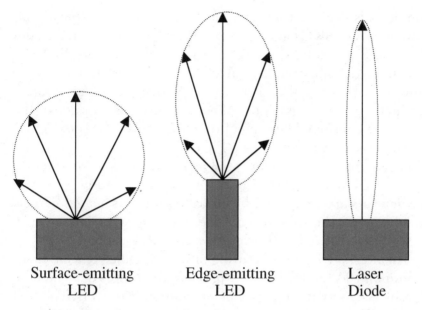

Surface-emitting   Edge-emitting   Laser
LED   LED   Diode

FIGURE 2-8   Light emission patterns for common semiconductor devices

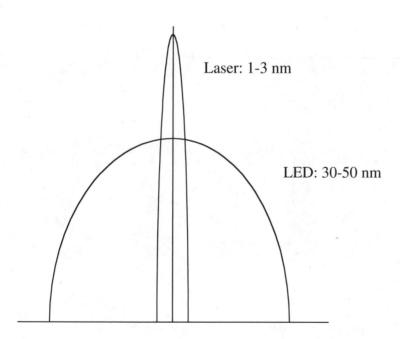

Laser: 1-3 nm

LED: 30-50 nm

FIGURE 2-9   Laser versus LED

compounds become the core. The compounds are typically silicon chloride (SiCl$_4$) and oxygen (O$_2$), which are reacted under heat to form a soft, sooty deposit of silicon dioxide (SiO$_2$) as shown in the illustration.[1] In practice, the SiCl$_4$ and O$_2$ are pumped into the fused silica tube as gases; the tube is heated, causing the sooty deposit to collect on the inside surface of the tube. The continued heating of the tube causes the soot to fuse into a glasslike substance (see Figure 2-10).

This process can be repeated as many times as required to create a graded refractive index, if required. Ultimately, once the deposits are complete, the entire assembly is heated fiercely which causes the tube to collapse, creating what is known in the optical fiber industry as a *preform*.

An alternative manufacturing process is called *outside vapor deposition* (OVD). In the OVD process, the soot is deposited on the surface of a rotating ceramic cylinder in two layers. The first layer is the soot that will become the core; the second layer becomes the cladding. Ultimately, the rod and soot are sintered[2] to create a preform. The ceramic is then removed, leaving behind the fused silica that will become the fiber.

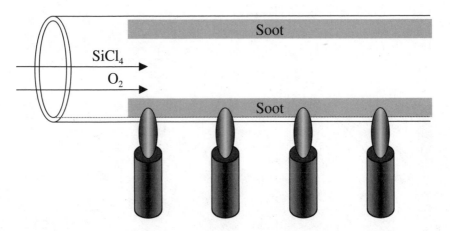

FIGURE 2-10   Fiber manufacture process

[1]In some cases, germanium chloride may also be added if the fiber is to be doped.

[2]A process in which a compound is heated to form a solid mass, but without melting.

A number of other techniques for creating the preforms are also used to create fiber, but these are the principal techniques in use today.

The next step is to convert the preform into optical fiber.

DRAWING THE FIBER

To make fiber from a preform, the preform is mounted at the top of a building called a *drawing tower* (see Figure 2-11). The bottom of the preform is heated until it has the consistency of taffy, at which time the soft glass is drawn down to form a thin fiber. When it strikes the cooler air outside the furnace, the fiber solidifies. Needless to say, the process is carefully managed to ensure that the thickness of the fiber is precise.

Other stages in the manufacturing process include monitoring processes to check the integrity of the product, a coating process that applies a protective layer, and a take-up stage where the fiber is wound onto reels for later assembly into cables of various types.

OPTICAL FIBER

Dozens of different types of fiber exist. Some of them are holdovers from previous generations of optical technology that

FIGURE 2-11   Drawing tower (courtesy of Lucent Technologies) Denmark

are still in use and represented the best efforts of technology available at the time. Others represent improvements on the general theme or specialized solutions to specific optical transmission challenges.

Generally speaking, two major types of fiber exist: *multimode,* which is the earliest form of optical fiber and is characterized by a large diameter central core, short distance capability, and low bandwidth; and *single mode,* which has a narrow core and is capable of greater distance and higher bandwidth. Varieties of each will be discussed in detail later in the book.

To understand the reason for and philosophy behind the various forms of fiber, it is first necessary to understand the issues that confront transmission engineers who design optical networks.

Optical fiber has a number of advantages over copper: it is lightweight, has enormous bandwidth potential, has significantly higher tensile strength, can support many simultaneous channels, and is immune to electromagnetic interference. It does, however, suffer from several disruptive problems that cannot be discounted. The first of these is *loss* or *attenuation*—the inevitable weakening of the transmitted signal over distance that has a direct analog in the copper world. Attenuation is typically the result of two subproperties, *scattering* and *absorption,* both of which have cumulative effects. The second is *dispersion,* which is the spreading of the transmitted signal and is analogous to noise.

Scattering

Scattering occurs because of impurities or irregularities in the physical makeup of the fiber itself. The best known form of scattering is called *Rayleigh Scattering*. It is caused by metal ions in the silica matrix and results in light rays being scattered in various directions as shown in Figure 2-12.

Rayleigh Scattering occurs most commonly around wavelengths of 1000 nm, and is responsible for as much as 90 percent of the total attenuation that occurs in modern optical

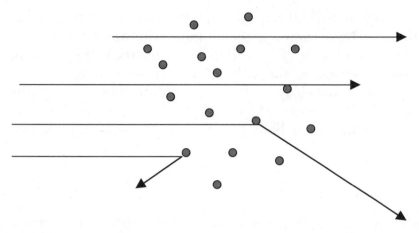

FIGURE 2-12   Scattering

systems. It occurs when the wavelengths of the light being transmitted are roughly the same size as the physical molecular structures within the silica matrix; thus, short wavelengths are affected by Rayleigh Scattering effects far more than long wavelengths. In fact, it is because of Rayleigh Scattering that the sky appears to be blue—the shorter (blue) wavelengths of light are scattered more than the longer wavelengths of light.

ABSORPTION

Absorption results from three factors: hydroxyl (OH^-; water) ions in the silica; impurities in the silica; and incompletely diminished residue from the manufacturing process. These impurities tend to absorb the energy of the transmitted signal and convert it to heat, resulting in an overall weakening of the signal. Hydroxyl absorption occurs at 1.25 and 1.39 µ; at 1.7 µ, the silica itself starts to absorb energy because of the natural resonance of silicon dioxide.

It is interesting to take a side road for a moment and examine hydroxyl absorption, because it is the basis for the proper functionality of a commonly used household device—the microwave oven. If you place a dry, glass-measuring cup in a microwave oven, you can cook it in the oven for an hour and it will barely be warm to the touch. Fill it with water, however, or simply wet it lightly, and it will rapidly reach the boiling point of water. This happens simply because microwave ovens generate energy at a wavelength that is effectively absorbed by the hydroxyl ions in water. The hydroxyl ions convert the energy to heat, and the water boils, the meat cooks, and the potatoes bake.

DISPERSION

As mentioned earlier, dispersion is the optical term for the spreading of the transmitted light pulse as it transits the fiber. It is a bandwidth-limiting phenomenon and comes in two forms: *multimode dispersion*, and *chromatic dispersion*. Chromatic dispersion is further subdivided into *material dispersion* and *waveguide dispersion*.

MULTIMODE DISPERSION

To understand multimode dispersion, it is first important to understand the concept of a *mode*. Figure 2-13 shows a fiber with a relatively wide core. Because of the width of the core, it enables light rays arriving from the source at a variety of angles (three in this case) to enter the fiber and be transmitted to the

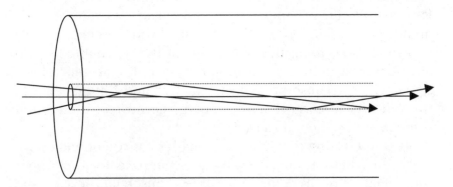

FIGURE 2-13   Multimode dispersion

receiver. Because of the different paths that each ray, or mode, will take, they will arrive at the receiver at different times, resulting in a dispersed signal.

Now consider the system shown in Figure 2-14. The core is much narrower, and only enables a single ray, or mode, to be sent down the fiber. This results in less end-to-end energy loss and avoids the dispersion problem that occurs in multimode installations.

CHROMATIC DISPERSION

The speed at which an optical signal travels down a fiber is absolutely dependent upon its wavelength. If the signal comprises multiple wavelengths, then the different wavelengths will travel at different speeds, resulting in an overall spreading or smearing of the signal. As discussed earlier, chromatic dispersion comprises two subcategories: material dispersion and waveguide dispersion.

MATERIAL DISPERSION

Simply put, material dispersion occurs because different wavelengths of light travel at different speeds through an optical fiber. To minimize it, two factors must be managed. The first of these is the number of wavelengths that make up the transmitted signal. An LED, for example, emits a rather broad range of wavelengths between 30 and 180 nm, whereas a laser emits a

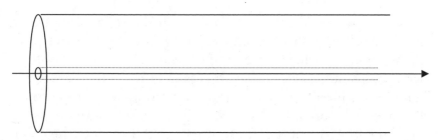

FIGURE 2-14   A single mode system

much narrower spectrum—typically less than five nm. Thus, a laser's output is far less prone to be seriously affected by material dispersion than the signal from an LED.

The second factor that affects the degree of material dispersion is a characteristic called the *center operating wavelength of the source signal*. In the vicinity of 850 nm, red, longer wavelengths travel faster than their shorter blue counterparts, but at 1550 nm, the situation's the opposite: blue wavelengths travel faster. Of course, a point at which the two meet and share a common minimum dispersion level does happen; it is in the range of 1310 nm, often referred to as the zero-dispersion wavelength. Clearly, this is an ideal place to transmit data signals, because dispersion effects are minimized here. As we will see later, however, other factors crop up that make this a less desirable transmission window than it appears.

Material dispersion is a particularly vexing problem in single-mode fibers.

WAVEGUIDE DISPERSION

Because the core and the cladding of a fiber have slightly different indices of refraction, the light that travels in the core moves slightly slower than the light that escapes into and travels in the cladding. This results in a dispersion effect that can be corrected by transmitting at specific wavelengths where material and waveguide dispersion actually cancel each other.

PUTTING IT ALL TOGETHER

So what does all of this have to do with the high-speed transmission of voice, video, and data? A lot, as it turns out. Understanding where attenuation and dispersion problems occur helps optical design engineers determine the best wavelengths at which to transmit information, taking into account distance, type of fiber, and other factors that can potentially affect the integrity of the transmitted signal. Consider the graph shown in Figure 2-15. It depicts the optical transmission

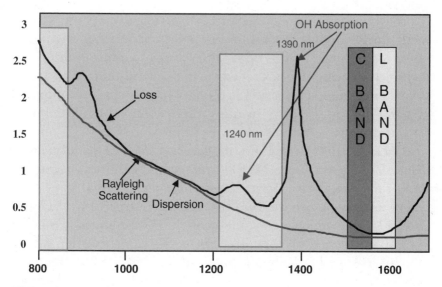

FIGURE 2-15   The optical domain

domain, as well as the areas where problems arise. Attenuation (dB/km) is shown on the Y-Axis; Wavelength (nm) is shown on the X-Axis.

First of all, note that four *transmission windows* appear in the diagram. The first one is at approximately 850 nm, the second at 1310 nm, a third at 1550 nm, and a fourth at 1625 nm, the last two labeled "C" and "L" band, respectively. The 850 nm band was the first to be used because of its adherence to the wavelength at which the original LED technology operated. The second window, at 1310 nm, enjoys low dispersion, as noted earlier. 1550 nm, the so-called C-Band, has emerged as an ideal wavelength for long-haul systems, although the relatively new L-Band has enjoyed some early success as the next effective operating window.

Notice also that Rayleigh Scattering is shown to occur at or around 1000 nm, although hydroxyl absorption by water occurs at 1240 and 1390 nm. Needless to say, network designers would be well served to avoid transmitting at any of the points on the graph where Rayleigh Scattering, high degrees of loss or hydroxyl absorption occur. Notice also that dispersion, shown by the lower line, is at a minimum point in the second window, while loss, shown by the upper line, drops to a minimum point in the third window. In fact, dispersion is minimized in traditional single-mode fiber at 1310 nm, while loss is at minimums at 1550 nm. So the obvious question becomes this: Which one do you want to minimize—loss, or dispersion?

Luckily, this choice no longer has to be made. Today, *dispersion-shifted fibers (DSF)* have become common. By modifying the manufacturing process, engineers were able to shift the point at which minimum dispersion occurs from 1310 nm to 1550 nm, causing it to coincide with the minimum loss point such that loss and dispersion occur at the same wavelength.

Unfortunately, although this fixed one problem, it created a new and potentially serious alternative problem. *Dense Wavelength Division Multiplexing* (DWDM) has become a mainstay technology for multiplying the available bandwidth in optical systems. When DWDM is deployed over dispersion-shifted fiber, serious nonlinearities occur at the zero dispersion point, discussed in detail later, which effectively destroy the DWDM signal. Specifically, a problem called *four-wave mixing* creates "sidebands" that interfere with the DWDM channels, destroying their integrity. In response, fiber manufacturers have created *non-zero dispersion-shifted fiber* (NZDSF) that lowers the dispersion point to near zero by making it occur just outside of the 1550 nm window. This eliminates the nonlinear four-wave mixing problem. These fiber options will be discussed in more detail later in this chapter.

FIBER NONLINEARITIES

A classic business quote, imminently applicable to the optical networking world, observes "in its success lie the seeds of its own destruction." As the marketplace clamors for longer transmission distances with minimal amplification, more wavelengths per fiber, higher bit rates, and increased signal power, a rather ugly collection of transmission impairments, known as *fiber nonlinearities,* rise to challenge attempts to make them happen. These impairments go far beyond the simple concerns brought about by loss and dispersion; they represent the performance wall against which optical networking is currently bumping.

THE POWER/REFRACTIVE INDEX PROBLEM

Two fundamental things cause the bulk of these nonlinearities. The first (and perhaps most critical) is the fact that the refractive index of the core of an optical fiber is directly dependent upon the power of the optical signal that is being transmitted through it. Because of this relationship, two actions can be taken to minimize the power-related problem. The first, obviously, is to minimize the transmitted power of the signal. This, however, has the downside of limiting the transmission distance and is a less-than-desirable option. The second, which is much more acceptable, is to maximize what is known as the fiber's effective area, a measure of the cross-sectional area of the fiber core, which carries the transmitted signal. By broadening the effective area of the fiber, it gathers more of the transmitted signal and thus reduces the need for an inordinately strong signal. Lucent's TrueWave® Fiber and Corning's Large Effective Area (LEAF®) fiber are examples of specially engineered products designed to overcome this problem.

The "special relationship" that exists between transmission power and the refractive index of the medium gives rise to four service-affecting optical nonlinearities: *self-phase modulation* (SPM), *cross-phase modulation* (XPM), *four-wave mixing* (FWM), and *intermodulation.*

SELF-PHASE MODULATION (SPM)

When self-phase modulation occurs, chromatic dispersion kicks in to create something of a technological double-whammy. As the pulse moves down the fiber, its leading edge increases the refractive index of the core, which causes a shift toward the blue end of the spectrum. The trailing edge, on the other hand, decreases the refractive index of the core, causing a shift toward the red end of the spectrum. This causes an overall spreading or smearing of the transmitted signal, a phenomenon known as *chirp*. It occurs in fiber systems that transmit a single pulse down the fiber and is proportional to the amount of chromatic dispersion in the fiber: the more chromatic dispersion, the more SPM. It is counteracted with the use of large, effective area fibers.

CROSS-PHASE MODULATION (XPM)

When multiple optical signals travel down the same fiber core, they both change the refractive index according to their own power levels. If they happen to cross, they will distort each other. Although XPM is similar to SPM, one significant difference occurs: while SPM is directly affected by chromatic dispersion, XPM is only minimally affected by it. Large, effective area fibers can reduce the impact of XPM.

FOUR-WAVE MIXING (FWM)

Four-wave mixing is the most serious of the power/refractive index-caused nonlinearities today because it has a catastrophic effect on DWDM-enhanced systems. Because the refractive index of fiber is nonlinear, and because multiple optical signals travel down the fiber in DWDM systems, a phenomenon known as *third-order distortion* occurs. Third-order distortion causes harmonics to be created in large numbers that have the annoying habit of occurring where the actual signals are, resulting in their obliteration. These harmonics tend to become numerous according to the equation

$$1/2(N^3-N^2)$$

where N is the number of signals. So if a DWDM system is transporting 16 channels, the total number of potentially destructive harmonics created would be 1,920.

Several things can reduce the impact of FWM. As dispersion in the fiber is reduced, the degree of four-wave mixing increases dramatically. In fact, it is *worst* at the zero-dispersion point. Thus, chromatic dispersion actually helps to reduce the effects of FWM. For this reason, fiber manufacturers sell *non-zero dispersion shifted fiber,* which, as described briefly earlier, moves the dispersion point to a point *near* the zero point, thus ensuring that a small amount of dispersion creeps in to protect against FWM problems.

Another factor that can minimize the impact of FWM is to widen the spacing between DWDM channels. This, of course, reduces the efficiency of the fiber by reducing the total number of available channels, and is therefore not a popular solution, particularly because the trend in the industry is to move toward narrower channel spacing as a way to increase the total number of available channels.

Finally, large effective area fibers tend to suffer less from the effects of FWM.

INTERMODULATION EFFECTS

In the same way that cross-phase modulation results from interference between multiple simultaneous signals, intermodulation causes secondary frequencies to be created that are cross-products of the original signals being transmitted. Large, effective area fibers can alleviate the symptoms of intermodulation.

SCATTERING PROBLEMS

Scattering within the silica matrix causes the second major impairment phenomenon. Two significant nonlinearities result: *Stimulated Brillouin Scattering* (SBS), and *Stimulated Raman Scattering* (SRS).

STIMULATED BRILLOUIN SCATTERING (SBS)

SBS is a power-related phenomenon. The power level of a transmitted optical signal remains below a certain threshold, usually on the order of three milliwatts—an annoyingly low level. The threshold is directly proportional to the fiber's effective area, and because dispersion-shifted fibers typically have smaller effective areas, they have lower thresholds. The threshold is also proportional to the width of the originating laser pulse: as the pulse gets wider, the threshold goes up. Thus, steps are often taken through a variety of techniques to artificially broaden the laser pulse. This can raise the threshold significantly, to as high as 40 milliwatts.

SBS is caused by the interaction of the optical signal moving down the fiber with the acoustic vibration of the silica matrix that makes up the fiber. As the silica matrix resonates, it causes some of the signal to be reflected back toward the source of the signal, resulting in noise, signal degradation and a reduction of overall bit rate of the system. As the power of the signal increases beyond the threshold, more of the signal is reflected, resulting in a multiplication of the initial problem.

It is interesting to note that Brillouin has two forms of scattering. When (sorry, a little more physics) electric fields that oscillate in time within an optical fiber interact with the natural acoustic resonance of the fiber material itself, the result is a tendency to backscatter light as it passes through the material. This is called *Brillouin Scattering*. If, however, the electric fields are caused by the optical signal itself, the signal is seen to cause the phenomenon; this is called *Stimulated Brillouin Scattering*.

To summarize: Because of backscattering, SBS reduces the amount of light that actually reaches the receiver and cause noise impairments. The problem increases quickly above the threshold, and has a more deleterious impact on longer wavelengths of light. One additional fact: in-line optical amplifiers (discussed later) add to the problem significantly. If four optical amplifiers appear along an optical span, the threshold will drop by a factor of four.

Solutions to SBS include the use of wider-pulse lasers and large effective area fibers.

Stimulated Raman Scattering (SRS)

Stimulated Raman Scattering (SRS) is something of a power-based crosstalk problem. In SRS, high-power, short wavelength channels tend to bleed power into longer wavelength, lower-power channels. It occurs when a light pulse moving down the fiber interacts with the crystalline matrix of the silica, causing the light to (1) be back-scattered and (2) shift the wavelength of the pulse slightly. Whereas SBS is a backward-scattering phenomenon, SRS is a two-way phenomenon, causing both back-scattering and a wavelength shift. The result is crosstalk between channels.

The good news is that SRS occurs at a much higher power level—close to a watt. Furthermore, it can be effectively reduced through the use of large effective area fibers.

AN ASIDE: OPTICAL AMPLIFICATION

As long as we are on the subject of Raman Scattering, we should introduce the concept of optical amplification. This may seem like a bit of a non sequitur, but it really isn't; true optical amplification actually uses a form of Raman Scattering to amplify the transmitted signal!

Traditional Amplification and Regeneration Techniques

In a traditional metallic analog environment, transmitted signals tend to weaken over distance. To overcome this problem, amplifiers are placed in the circuit periodically to raise the power level of the signal. This technique has a problem, however: in addition to amplifying the signal, amplifiers also amplify whatever cumulative noise has been picked up by the signal during its transit across the network. Over time, it becomes difficult for a receiver to discriminate between the actual signal and the noise embedded in the signal. Extraordinarily complex recovery mechanisms are required to discriminate between wheat and chaff.

In digital systems, *regenerators* are used to not only amplify the signal, but to also remove any extraneous noise that has been picked up along the way. Thus, digital regeneration is a far more effective signal recovery methodology than simple amplification.

Even though signals propagate significantly farther in optical fiber than they do in copper facilities, they are still eventually attenuated to the point that they must be regenerated. In a traditional installation, the optical signal is received by a receiver circuit, converted to its electrical analog, regenerated, converted back to an optical signal, and transmitted onward upon the next fiber segment (see Figure 2-16). This optical-to-electrical-to-optical conversion process is costly, complex, and time consuming. However, it is proving to be far less necessary as an amplification technique than it used to be because of true optical amplification that has recently become commercially feasible. Please note that optical amplifiers *do not* regenerate signals; they merely amplify. Regenerators are still required, albeit far less frequently.

Optical amplifiers represent one of the technological leading edges of data networking. Instead of the O-E-O process described previously, optical amplifiers receive the optical signal, amplify it as an optical signal, and then retransmit it as an optical signal—no electrical conversion is required (see Figure 2-17). Like their electrical counterparts, however, they also amplify the noise; at some point signal regeneration is required.

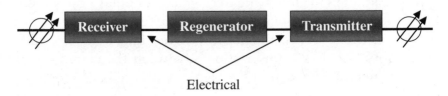

Electrical

FIGURE 2-16   Optoelectrical amplification

FIGURE 2-17   Optical amplification

OPTICAL AMPLIFIERS: HOW THEY WORK

It was only a matter of time before all-optical amplifiers became a reality. It makes intuitively clear sense that a solution that eliminates the O-E-O process would be a good one. Optical amplification is that solution.

You will recall that Stimulated Raman Scattering is a fiber nonlinearity characterized by high-energy channels pumping power into low-energy channels. What if that phenomenon could be harnessed as a way to amplify optical signals that have weakened over distance?

Optical amplifiers are actually rather simple devices that as a result tend to be extremely reliable. As Figure 2-18 illustrates, the optical amplifier comprises the following: an input fiber, carrying the weakened signal that is to be amplified; a pair of optical isolators; a coil of doped fiber; a pump laser; and the output fiber that now carries the amplified signal.

The coil of doped fiber lies at the heart of the optical amplifier's functionality. Doping is simply the process of embedding some kind of functional impurity in the silica matrix of the fiber when it is manufactured. In optical amplifiers, this impurity is more often than not an element called *erbium*. Its role will become clear in just a moment.

The pump laser shown in the upper left corner of Figure 2-18 generates a light signal at 980 nm in the *opposite direction* than the actual signal flows. As it turns out, erbium becomes atomically excited when it is struck by light at that wavelength. When an atom is excited by pumped energy, it jumps to a higher

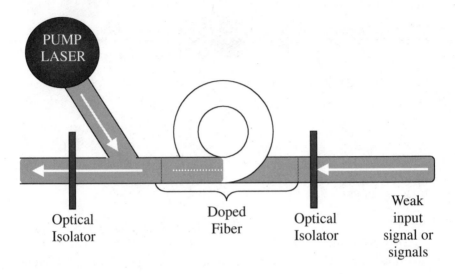

FIGURE 2-18   Optical Amplifier

energy level, then falls back down, during which it gives off a photon at a certain wavelength. When erbium is excited by light at 980 nm, it emits photons at 1550 nm—coincidentally the window in which multichannel optical systems operate. So, when the weak, transmitted signal reaches the coil of erbium-doped fiber, the erbium atoms, now excited by the energy from the pump laser, bleed power into the weak signal at precisely the right energy level, causing a generalized amplification of the transmitted signal. The optical isolators serve to prevent errant light from backscattering into the system, creating noise. Of course, *erbium-doped fiber amplifiers* (EDFAs) are highly proletariat in nature; they amplify anything, including the noise that the signal may have picked up. Therefore, a need still exists at some point along the path of long-haul systems for regeneration, although far less frequently than in traditional copper systems.[3]

---

[3]I have a friend who claims he's going to make a fortune selling erbium body cream. You go into a dark room, smear it all over, and turn on the light . . .

Other Amplification Options

At least two other amplification techniques in addition to EDFAs exist that have recently come into favor. The first of these is called *Raman amplification,* which is similar to EDFA in the sense that it relies on Raman effects to do its task, but different for other rather substantial reasons. In Raman amplification, the signal beam travels down the fiber alongside a rather powerful pump beam, which excites atoms in the silica matrix that in turn emit photons that amplify the signal. The advantage of Raman amplification is that it requires no special doping: erbium is not necessary. Instead, the silica gives off the necessary amplification. In this case, the fiber itself becomes the amplifier.

Raman amplifiers require a significantly high-power pump beam (about a watt, although some systems have been able to reduce the required power to 750 mw or less) and even at high levels the power gain is relatively low. Their advantage, however, is that their induced gain is distributed across the entire optical span. Furthermore, it will operate within a relatively wide range of wavelengths, including 1310 and 1550 nm, two of the most popular and effective transmission windows.

Semiconductor lasers have also been deployed as optical amplification devices in some installations. In semiconductor optical amplifiers, the weakened optical signal is pumped into the ingress edge of a semiconductor optical amplifier as shown in Figure 2-19. The active layer of the semiconductor substrate amplifies the signal and regenerates it on the other side. The primary downside to these devices is their size—they are small, and their light-collecting ability is therefore somewhat limited. A typical single-mode fiber generates an intense spot of light that is roughly 10 microns in diameter. The point upon which that light impinges upon the semiconductor amplifier is less than a micron in diameter, meaning that a lot of the light is lost. Other problems also crop up including polarization issues, reflection, and variable gain. As a result, these devices are not in widespread use; EDFAs and Raman amplification techniques are far more common.

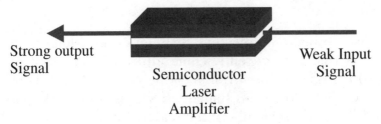

Strong output Signal

Semiconductor Laser Amplifier

Weak Input Signal

FIGURE 2-19   A semiconductor laser amplifier

PULLING IT ALL TOGETHER

So why do we care about these challenges, problems, and non-linearities? Because they have a direct effect on the degree to which we can transmit signals through optical media. There is no question that the fiber is orders of magnitude better than copper as a transmission medium for broadband signals, but it does have limitations that cannot be ignored. In the same way that data transmitted over copper networks suffers impairments from cumulative noise and signal deterioration, so too do optical signals. As the demands for higher bandwidth and greater degrees of channelization grow, these impairments must be carefully managed to prevent them from having an inordinately large impact on the systems in which they occur. The good news is that they are well understood; and optical network engineers have developed good measurement tools to detect them and systems to control them.

OPTICAL RECEIVERS

So far, we have discussed the sources of light, including LEDs and laser diodes. We have briefly described the various flavors of optical fiber and the problems they encounter as transmission media. Now, we turn our attention to the devices that receive the transmitted signal.

The receive devices used in optical networks have a single responsibility: to capture the transmitted optical signal and convert it into an electrical signal that can then be processed by the end equipment. There may also be various stages of amplification to ensure that the signal is strong enough to be acted upon, and demodulation circuitry, which recreates the originally transmitted electronic signal.

Think for a moment about the term *semiconductor*. A semiconductor is a compound that—well, only semi-conducts. It sits in the gray area between conductors and insulators, and must be somehow induced to conduct current. The optical receivers that are commonly used in optical networks are semiconductors themselves. Let's take a moment to describe how they work. Sorry, we must descend once again into the bowels of physics to do this.

Photosensitive semiconductors, which are silicon-based, typically consist of three functional layers as shown in Figure 2-20: a negative region, a positive region, and a junction region. Photodetectors are said to be *reverse-biased* because the negative charges ("electrons") and the positive charges ("holes") are prevented from migrating into the center junction region, thus

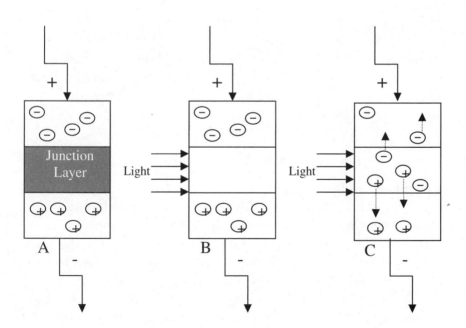

FIGURE 2-20   A semiconductor receiver

preventing the flow of current through the semiconductor from one active layer to the other (A). This changes (B) when light of a specific wavelength strikes the photosensitive junction layer, causing the creation of electron-hole pairs in the junction layer. This results in an overall flow of current that is proportional to the intensity of the light striking the junction layer.

Different substances can be used as photodetectors, including *silicon (Si)*, *germanium (Ge)*, *gallium arsenide (GaAs)*, and *indium-gallium arsenide (InGaAs)*, to name a few. They are selected based upon the operating wavelength in which they will be used, because their sensitivity to light varies according to the information contained in the following table.

SUBSTANCE	OPERATING WAVELENGTH (NM)
silicon	400-1100
germanium	800-1600
gallium arsenide	400-1000
Indium-gallium arsenide	400-1700

PHOTODETECTOR TYPES

Although many different types of photosensitive devices exist, two are used most commonly as photodetectors in modern networks: *positive-intrinsic-negative* (PIN) *photodiodes*, and *avalanche photodiodes* (APDs).

POSITIVE-INTRINSIC-NEGATIVE (PIN) PHOTODIODES

PIN photodiodes are similar to the device described above in the general discussion of photosensitive semiconductors. Reverse biasing the junction region of the device prevents current flow until light at a specific wavelength strikes the substance, creating electron-hole pairs and allowing current to flow across the three-layer interface in proportion to the intensity of the incident light. Although they are not the most sensitive devices available for the purpose of photodetection, they are perfectly adequate for the requirements of most optical systems. In cases where they are not considered sensitive enough for high-performance systems, they can be coupled with a pre-amplifier to increase the overall sensitivity.

Avalanche Photodiodes (APD)

Avalanche photodiodes work as optical signal amplifiers (see Figure 2-21). They use a strong electric field to perform what is known as *avalanche multiplication*. In an APD, the electric field causes current accelerations such that the atoms in the semiconductor matrix get excited and create, in effect, an "avalanche" of current to occur. The good news is that the amplification effect can be as much as 30 to 100 times the original signal; the bad news is that the effect is not altogether linear and can create noise. APDs are sensitive to temperature and require a significant voltage to operate them—30 to 300

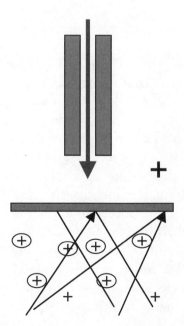

FIGURE 2-21   Avalanche photodiode

volts depending on the device. However, they are popular for broadband systems and work well in the gigabit range.

We have now discussed transmitters, fiber media, and receivers. In the next section, we examine the fibers themselves, and how they have been carefully designed to serve as solutions for a wide variety of networking challenges and to forestall the impact of the nonlinearities described in this section.

OPTICAL FIBER

As was mentioned briefly in a prior section, fiber has evolved over the years in a variety of ways to accommodate both the changing requirements of the customer community and the technological challenges that emerged as the demand for bandwidth climbed precipitously. These changes came in the form of various forms of fiber that presented different behavior characteristics to the market.

MULTIMODE FIBER

The first of these was *multimode fiber,* which came in a variety of different forms. Multimode fiber bears that name because it allows more than a single mode or ray of light to be carried through the fiber simultaneously due to the relatively wide core diameter that characterizes the fiber (refer to Figures 2-13 and 2-14). And although the dispersion that potentially results from this phenomenon can be a problem, there were advantages to the use of multimode fiber. For one thing, it is far easier to couple the relatively wide and forgiving end of a multimode fiber to a light source than that of the much narrower single mode fiber. It is also significantly less expensive and relies on LEDs and inexpensive receivers rather than the more expensive laser diodes and ultra-sensitive receiver devices. However, advancements in technology have caused the use of multimode fiber to fall out of favor. Single-mode is far more commonly used today.

Multimode fiber is manufactured in two forms: *step-index fiber,* and *graded-index fiber.* We will examine each in turn.

MULTIMODE STEP-INDEX FIBER

In step-index fiber, the index of refraction of the core is *slightly* higher than the index of refraction of the cladding (see Figure 2-22). Remember that the higher the refractive index, the slower the signal travels through the medium. Thus, in step-index fiber, any light that escapes into the cladding because it enters the core at too oblique an angle will actually travel slightly faster in the cladding (assuming it does not escape altogether) than it would if it traveled in the core. Of course, any rays that are reflected repeatedly as they traverse the core also take longer to reach the receiver, resulting in a dispersed signal that causes problems for the receiver at the other end. Clearly, this phenomenon is undesirable; for that reason, graded-index fiber was developed.

MULTIMODE GRADED-INDEX FIBER

Because of the dispersion that is inherent in the use of step-index fiber, optical engineers created graded index fiber as a way to overcome the signal degradation that occurred.

In graded-index fiber, the refractive index of the core actually decreases from the center of the fiber outward, as shown in Figure 2-23. In other words, the refractive index at the center of the core is higher than the refractive index at the edge of the

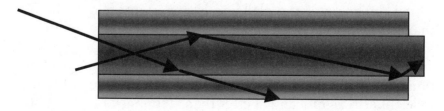

FIGURE 2-22 Multimode step-index fiber

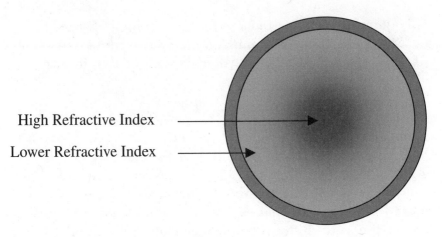

High Refractive Index

Lower Refractive Index

FIGURE 2-23   Multimode graded-index fiber

core. The result of this rather clever design is that as light enters the core at multiple angles and travels from the center of the core outward, it is actually accelerated at the edge and slowed down near the center, causing most of the light to arrive at roughly the same time. Thus, graded-index fiber helps to overcome the dispersion problems associated with step-index multimode fiber. Light that enters this type of fiber does not travel in a straight line, but rather follows a parabolic path as shown in Figure 2-24, with all rays arriving at the receiver at more or less the same time.

Graded-index fiber typically has a core diameter of 50 to 62.5 microns, with a cladding diameter of 125 microns. Some variations exist; for example, at least one form of multimode graded-index appears with a core diameter of 85 microns, somewhat larger than those described above. Furthermore, the actual thickness of the cladding is important: if it is thinner than 20 microns, light begins to seep out, causing additional problems for signal propagation.

Graded-index fiber was commonly used in telecommunications applications until the late 1980s. Even though graded-index fiber is significantly better than step-index fiber, it is still multimode fiber and does not eliminate the problems inherent

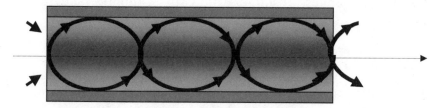

FIGURE 2-24   Signal propagation in graded-index fiber

in being multimode. Thus was born the next generation of optical fiber: single-mode.

SINGLE-MODE FIBER

An interesting mental conundrum crops up with the introduction of single-mode fiber. The core of single-mode fiber is significantly narrower than the core of multimode fiber. Because it is narrower, it would seem that its ability to carry information would be reduced due to limited light-gathering ability. This, of course, is not the case. As its name implies, it enables a single mode or ray of light to propagate down the fiber core, thus eliminating the intermodal dispersion problems that plague multimode fibers. In reality, single-mode fiber is a stepped-index design, because the core's refractive index is slightly higher than that of the cladding. It has become the de facto standard for optical transmission systems, and takes on many forms depending on the specific application within which it will be used.

Most single-mode fiber has an extremely narrow core diameter on the order of seven to nine microns, and a cladding diameter of 125 microns. The advantage of this design is that it only allows a single mode to propagate; the downside, however, is the difficulty involved in working with it. The core must be coupled directly to the light source and the receiver in order to make the system as effective as possible. Given that the core is approximately one-sixth the diameter of a human hair, the mechanical process through which this coupling takes place becomes Herculean.

SINGLE-MODE FIBER DESIGNS

The reader will recall that we spent a considerable amount of time discussing the many different forms of transmission impairments (nonlinearities) that challenge optical systems. Loss and dispersion are the key contributing factors in most cases, and do in fact cause serious problems in high-speed systems. The good news is that optical engineers have done yeoman's work creating a wide variety of single-mode fibers that address most of the nonlinearities.

Since its introduction in the early 1980s, single-mode fiber has undergone a series of evolutionary phases in concert with the changing demands of the bandwidth marketplace. The first variety of single-mode fiber to enter the market was called *non-dispersion-shifted fiber* (NDSF). Designed to operate in the 1310 nm second window (refer to Figure 2-15), dispersion in these fibers was close to zero at that wavelength. As a result, it offered high bandwidth and low dispersion. Unfortunately, it was soon the victim of its own success. As demand for high-bandwidth transport grew, a third window was created at 1550 nm for single-mode fiber transmission. It provided attenuation levels that were less than half those measured at 1310 nm, but unfortunately was plagued with significant dispersion. Because the bulk of all installed fiber was NDSF, the only solution available to transmission designers was to narrow the linewidth of the lasers employed in these systems and to make them more powerful. Unfortunately, increasing the power and reducing the laser linewidth is expensive, so another solution soon emerged.

DISPERSION-SHIFTED FIBER (DSF)

One solution that emerged was dispersion-shifted fiber. With DSF, the minimum dispersion point is mechanically shifted from 1310 nm to 1550 nm by modifying the design of the actual fiber so that waveguide dispersion is increased. The reader will recall that waveguide dispersion is a form of chromatic dispersion that occurs because the light travels at different speeds in the core and cladding.

One technique for building DSF (sometimes called "zero dispersion-shifted fiber") is to actually build a fiber of multiple layers, as shown in Figure 2-25.

In this design, the core has the highest index of refraction that changes gradually from the center outward until it equals the refractive index of the outer cladding. An inner cladding layer surrounds the inner core, which is in turn surrounded by an outer core. This design works well for single wavelength systems, but experiences serious signal degradation when multiple wavelengths are transmitted, as in DWDM systems. Four-wave mixing, described earlier, becomes a serious impediment to clean transmission. Given that multiple wavelength systems are fast becoming the norm today, the single wavelength limit is a show-stopper. The result was a relatively simple and elegant set of solutions.

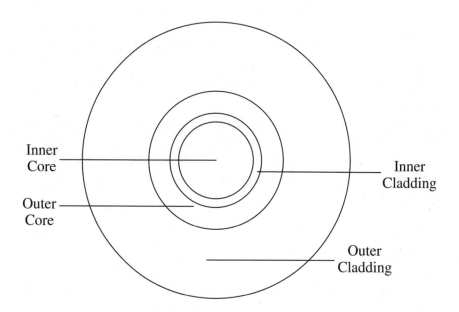

FIGURE 2-25   Dispersion-shifted fiber

The first of these was to maximize the effective area of the fiber, as described earlier. Lucent's TrueWave® Fiber and Corning's LEAF® Fiber are examples of this. Because the overall power of the optical signal(s) being carried by the fiber is distributed across a broader cross-section, the nonlinear performance problems are less pronounced.

The second technique was to eliminate or at least substantially reduce the absorption peaks in the fiber performance graph so that the second and third transmission windows merge into a single larger window, thus allowing for the creation of the fourth window described earlier which operates between 1565 and 1625 nm—the so-called L-Band.

Finally, the third solution came with the development of NZDSF. NZDSF shifts the minimum dispersion point so that it is *close* to the zero point, but not actually *at* it. This prevents the nonlinear problems that occur at the zero point.

WHY DO WE CARE?

It is always good to go back and review why we care about such things as dispersion-shifting and absorption issues. Remember that the key to keeping the cost of network down is to reduce maintenance and the need to add hardware or additional fiber when bandwidth gets tight. DWDM, discussed in detail in the next chapter, offers an elegant and relatively simple solution to the problem of the cost of bandwidth. However, its use is not without cost. Multi-wavelength systems will not operate effectively over dispersion-shifted fiber because of dramatic nonlinearities, so if DWDM is to be used, non-zero dispersion-shifted fiber must be deployed.

SUMMARY

In this section we have examined the history of optical technology and the technology itself, focusing on the three key components within an optical network: the light emitter, the transport medium, and the receiver. We also discussed the various forms of transmission impairment that can occur in optical systems, and the steps that have been taken to overcome them.

The result of all this is that optical fiber, once heralded as a near-technological miracle because it only lost 99 percent of its signal strength *when transmitted over an entire kilometer,* has become the standard medium for transmission of high-bandwidth signals over great distances. Optical amplification now serves as an augmentation to traditional regenerated systems, enabling the elimination of the optical-to-electrical conversion that must take place in copper systems. The result of all this is an extremely efficient transmission system that has the ability to play a role in virtually any network design in existence today.

In the next section, we will examine corollary technologies that add capability, richness, and application specificity to optical transmission including fiber cable design, DWDM, submarine systems, and optical switching and routing technologies.

PART THREE

COROLLARY TECHNOLOGIES

INTRODUCTION

In this section, we examine the supporting technologies that convert optical transmission from an intriguing laboratory experiment into a powerful component of the networking domain. These technologies include optical cabling assemblies, *Dense Wavelength Division Multiplexing* (DWDM), optical switching and routing technologies, and network management.

Keep in mind that optical transmission has been part of the technological pantheon available to the access and transport providers for nearly 30 years. However, it has only been recently that complementary technologies have become available that not only expand optical's reach, but also enrich and augment it to make it much more than a high-speed transport mechanism. SONET and SDH were the first major advances in optical technology, but as was discussed earlier, they are now being viewed as legacy technologies in many ways, and although still broadly installed, are not considered to be "service-rich" enough for the growing demands of the customer. They are also considered to be far too overhead-intensive to meet the growing demand for efficient, overhead-light protocol structures. Remember the key demands placed on the network by the market today and the drivers behind them:

- The need to create routes on demand between individual users as well as between disparate work groups, in response to the market shying away from dedicated, costly facilities.
- Guaranteed interoperability between disparate protocols.
- Universal, seamless connectivity between far-flung corporate locations.
- Optimum utilization of network bandwidth through the appropriate use of intelligent prioritization and routing techniques.
- Traffic aggregation for wide area transport to ensure efficient use of network bandwidth.
- Granular quality of service control through effective policy and queue management techniques.
- Growing deployment of high-speed access technologies such as DSL, cable modems, wireless local loop, and satellite connectivity.

When we examine these requirements—"unlimited" bandwidth on demand, interoperability, seamless connectivity, QoS, and support for high-speed access—optical solutions quickly rise to the top of the technological heap as optimum answers. The combination of blindingly high-speed access and transport, intelligent all-optical switching, multichannel transport, and effective, proactive network management combine into a powerful set of capabilities that will effectively address most, if not all, of these concerns.

OPTICAL CABLE ASSEMBLIES

So far, we have discussed optical fiber as a technology breakthrough that makes possible the transmission of extremely high bandwidth services across an impossibly small diameter medium. However, much more is involved. Once the silica preform has been created, suspended in the heated drawing tower, and drawn into optical fiber, the fiber itself must be cushioned and protected for installation. This is done by assembling one or more fibers into a cable that provides not only optical transport but cushioning, insulation, protection from undue bending (that can lead to loss), pulling strength, and other more specialized functions that are specific to each cable type. Optical fiber is used in a wide variety of environments, some of them hostile such as in submarine applications. These environmental variables must be taken into account when selecting the appropriate cable type for a network.

The environments in which optical cable is installed vary greatly, and fall into three main categories: inside installations; outside installations; and special purpose installations.

INSIDE INSTALLATIONS

Inside installations, as the name implies, are environments typically found inside office buildings, data centers, or sealed (and therefore environmentally sound) plenum vaults between buildings in a campus environment. At the shortest end of the scale, we find fiber installed as interconnect media within devices such as computers, switches, and large-scale routers. At the next level, fiber is often used for backbone transport within a workspace, such as the floor of a factory or within a call center.

More and more, fiber is being installed in riser ducts to provide backbone connectivity and transport in multi-tenant buildings. For example, a high-rise office building housing many different tenants requires voice, data, and video connectivity. And although each company's specific requirements may vary, they all need varying degrees of bandwidth and service levels (see the previous list). A fiber backbone that in turn interconnects to the optical network outside the building provides an ideal mechanism for "future-proofing" the facility's in-building backbone network.

OUTSIDE INSTALLATIONS

Outside installations must endure harsher environmental treatment than their inside counterparts, and must therefore be hardened against these challenges. At the least hostile end of the spectrum are the corporate campus installations used to provide connectivity between multiple buildings for interconnecting local area networks, videoconferencing facilities, and other corporate network resources. They are usually transported via dedicated ducts, often owned by the corporation, and are therefore relatively well protected against the harshness of the elements.

For longer distance installations, the fiber cable is often buried in plastic conduit (usually, but not always, orange). The conduit is installed either by digging a trench and laying the cable with other facilities in the trench, or by burrowing a narrow horizontal tunnel, usually slightly more than an inch in diameter, with a high-pressure water jet and pushing the conduit through the tunnel. This avoids the disruptive need to dig; the high water pressure is strong enough to burrow through anything short of granite. In some cases, specially hardened cable can be buried directly in the ground without protective conduit.

Cables are also installed as aerial facilities, lashed to strength members to support them. Because of wind force, they must be able to support a certain minimum amount of wind shear because of the stretching that they will often undergo.

SPECIAL PURPOSE CABLES

Special purpose cables are used for—well, special purpose applications. The most common of these are the cables used for submarine installations that must withstand abrasion from the geography of the sea floor, the high pressure of the environment, even shark attack. They must also provide power for the amplifiers and regenerators that are installed periodically along the route.

Other special purpose cables include those designed to withstand harsh chemical environments and high temperatures. They may be covered in a special, environment-resistant plastic sheath or in some cases, armored. For weight support purposes, some optical cables are designed in a "figure-8" configuration;

the upper half of the "8" is a steel support member and the lower half is the actual fiber cable. We will examine these in detail later.

FIBER CABLE ARCHITECTURES

Generally speaking, fiber cables are constructed in two broad categories: loose-tube cables, and tight-buffer cables. In a loose-tube cable, the fibers themselves are housed within a conduit but are not tightly packed—instead, they are allowed to "float" within the conduit. This enables them to move freely in response to movement of the cable itself, realigning themselves as required to prevent kinking and the problems that arise from stress and microbending. Microbends occur when the fibers are bent far enough to exceed design parameters, resulting in loss and dispersion from the refractive effects of the bending.

In many cases, the core of the conduit is filled with a water-absorbing gel to both cushion the fibers and prevent the damaging effects of water entry.

As Figure 3-1 illustrates, a loose-tube fiber cable is rather complex in its construction. A dielectric or steel central mem-

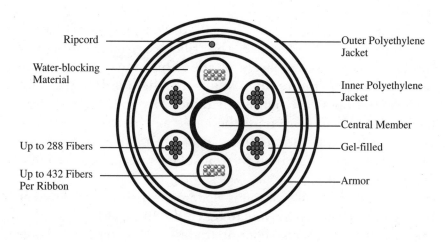

FIGURE 3-1 Loose-tube fiber cable

ber that provides strength and rigidity is surrounded helically by as many as 12 color-coded buffer tubes, each of which contains optical fibers. The tubes are often filled with a special gel to prevent the entrance of water and to cushion the fibers within. The fibers may be loose in the buffer tubes, in which case there may be as many as 288 of them in the cable (12 per tube); or, if installed as a ribbon, may contain as many as 432 fibers (18 per tube). Ribbons comprise multiple fibers that have been bonded into a ribbon-like assembly that makes the splicing of multiple fibers easier than splicing individual fibers. It enables very high fiber count density; however, ribbon cables can kink easily and must be handled carefully during installation and maintenance.

As always, the name of the game in fiber cable construction is cost-reduction. Anything that can be done to make the product more cost-effective is a good thing. The more fibers that can be functionally installed within the sheathing of a cable intended for long-distance service, the more cost-effective it is.

Surrounding the buffer tubes is a special water-absorbing material that adds further protection against flooding. The buffer tubes are then encased in an inner polyethylene jacket, followed by a protective armor cover, if warranted, to prevent the cable from being crushed or damaged by chewing animals such as squirrels. The armor is then covered with an outer polyethylene jacket.

Loose-tube designs are typically used in long-term, wide-area applications. Because the fibers are not physically "bonded" to the outer layers of the cable, they are far more resistant to temperature fluctuations than tight-buffered cables, because the fibers are tightly surrounded by and in direct contact with the outer layers of the cable assembly. Temperature changes can cause fluctuations in the refractive index of the fibers themselves, and therefore must be controlled to the highest possible degree.

A typical cable also contains ripcords, which are used to tear through the protective layers to reach the fiber within for splicing purposes. It may also contain traditional copper twisted pair and/or power leads for amplifiers. In some cases it may be bonded to a steel messenger cable in what is known as a "figure-8" configuration, as shown in Figure 3-2. The figure-8 is a self-supporting cable designed for aerial use in extreme conditions such as heavy wind and ice loading. The upper half of the "8" is an asphalt-filled steel cable; the lower half is the optical cable assembly.

Other options in cable manufacturing address specific environmental conditions under which the cable may have to operate. For example, the outer sheath can be made from a variety of materials to make the cable resistant to harsh chemical environments, fire, abrasion, cutting, pulling through conduits, cracking, cold, ozone, and the inevitable chewing effects of rodents in outside installations.

The alternative form of fiber cable construction is called *tight-buffer construction*. In tight-buffer assemblies, the fibers are locked into place within the cable and not allowed to move freely as they do in loose buffer cables. The fibers are actually surrounded by a soft, moldable plastic sheath that locks them into place. A harder, more rigid layer for protection then surrounds the soft plastic layer. Tight-buffer cables are often

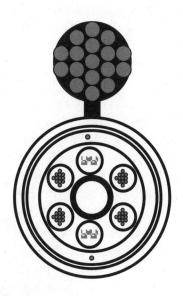

FIGURE 3-2   Figure-8 cable

smaller in diameter than loose-tube cables, and are therefore popular for use in inside installations where plenum vault and conduit space is at a premium.

THE SPECIAL CASE OF SUBMARINE CABLES

Cable & Wireless laid the first submarine cable (copper, of course, not optical) in 1856 between Newfoundland and Ireland. It provided telegraph connectivity for many years, before being decommissioned in the mid-1920s. It provided superb service, and served as a bellwether for what was to come.

Others followed. In 1943, the UK installed the first repeatered undersea cable between Port Erin in the extreme west and Holyhead. In North America, the first cable became operational between Havana, Cuba and Key West in 1950, while the first *transatlantic cable* (TAT-1), which was coaxial, became operational in 1956 between the UK and Canada. It carried 36 phone circuits and had repeaters approximately every 50 miles. TAT-7, which became operational in 1983 and is still in use, is repeatered every five miles and was the last coaxial cable put into service.

Needless to say, optical technology soon caught the attention of Bell Laboratories as an alternative transport solution for voice and data traffic. By the mid-1970s, they were working at a fast and furious pace to perfect optical transmission for use in TAT-8, the first transoceanic cable based on optical fiber. Their design called for three pairs of single mode fiber operating at 1300 nm, two live pairs and a third serving as a spare. The cable would transport voice and data at 278 Mbps on each pair, for an aggregate data rate in excess of 550 Mbps. Using voice compression technology, it permitted the simultaneous transport of 40,000 voice calls. TAT-8 called for repeaters to be installed every 30 miles, and to have mean-time-between-failure measurements that would guarantee a submerged lifespan of 25 years and the need for no more than two recoveries of the cable during that time for repairs. The cable became operational in 1988, and ushered in a whole new era of optical transmission, redefining the concept of "long-haul transport."

In the early 1990s, the third window of transport opened at 1550 nm, making it possible to employ EDFA technology and to increase the overall data rates of installed cables. TAT-9 and TAT-10 offered 565 Mbps of bandwidth and repeater spacing well in excess of 60 miles. TAT-12 and TAT-13, a remarkable optical loop connecting the USA, the UK and France, added EDFA technology and offered 10 Gbps of overall bandwidth. More recent cables offer optical add-drop multiplexing and enormous bandwidth—640 Gbps in the case of Project Oxygen. Oxygen is an ambitious $14 billion project funded by venture capital and intended to interconnect the world with flat-rate, distance-insensitive transport. Lucent Technologies and Corning have been selected as major suppliers of switching equipment and non-zero dispersion-shifted fiber, and although the project is not proceeding as quickly as its supporters would like, it is moving forward. The first leg is expected to be complete in the first half of 2001.

Like *Fiber Link Around the Globe* (FLAG), which interconnects Japan with the UK, Oxygen's intent is to provide a cost-effective solution to the perceived lack of global bandwidth. However, given the degree to which companies like Qwest and Global Crossing have deployed bandwidth, that lack may be more imagination than reality. Say's Law, of course, observing that supply creates its own demand, would indicate that any perceived fiber glut is purely momentary.

According to consultancy KMI, enough submarine optical capacity will be installed by 2003 to provide transport for the equivalent of 800 million simultaneous telephone calls. Furthermore, they observe that more than 70 percent additional fiber will be installed in the next five years than what has been installed since fiber was first used on the TAT-8 (the first transatlantic fiber cable), and TPC-3 cable projects in the 1980s. This will equate to approximately $23 billion in installed fiber in the Pacific basin alone, with more than $50 billion total invested by 2003.

FIBER INSTALLATION TECHNIQUES

When fiber is to be installed relatively close to shore, it is typically armored to prevent potential damage from ship anchors,
dredges, and trawlers. In deeper water, the extensive armoring
is not required. In fact, it is difficult to work with cable in deep
water because it is heavy and can break from its own weight
while being installed on the seabed.

When fiber is installed close to the shore and is used to
interconnect cities along a shoreline as shown in Figure 3-3,
the design is called *festooning*. In a festooned system, the fiber
is laid in great loops offshore, usually no more than 150 mile
long. The loops come ashore at cities where service is required;
amplifiers are usually installed at the landing points as well. In
areas where concern for the stability of the region and the
potential for damage to the cable system is real, a different

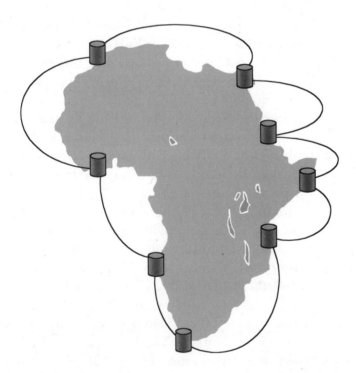

FIGURE 3-3   Festooning

technique is used. In those cases, optical spurs are installed as shown in Figure 3-4. These optical spurs do not provide access to the entire fiber; instead, they only bring ashore specific wavelengths selected for use by the region in question. A disruption of the cable at its landfall in this situation would only disrupt the signal going to that particular landfall point, but would not affect the rest of the cable. In these installations, the trunk (main ring) is usually about 1,000 miles offshore.

Submarine cables dwell in one of the harshest and most abusive environments possible. They must withstand high pressure, abrasion, attacks from biting animals such as sharks, unexploded ordinance from past wars, and damage from dredging activities and inadvertently dropped boat anchors.

When submarine cables are installed in shallow water, they must be protected from damage because of boating activity, and are therefore often buried. They are also less subject to the

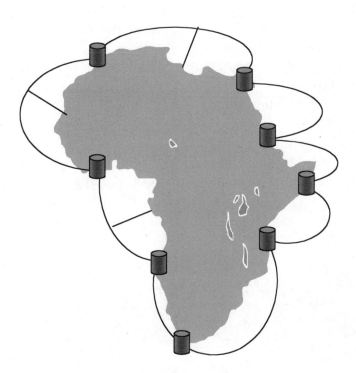

FIGURE 3-4   Optical spurs

rigors of high-pressure, and for shallow applications typically comprise nothing more than sealed versions of terrestrial cable with armor added to prevent crushing. For deepwater applications, the cable assembly is far more elaborate, as shown in Figure 3-5. Deep water cables are not typically subjected to abuse from shipping activity because of their depth, and therefore do not need to be buried. They are subject to leakage, earth movement, and biting, however, and must therefore be carefully protected as a result. Figure 3-5 shows the many layers that are often found in submarine assemblies, including a nylon outer covering, various polypropylene layers, several layers of steel wire armor, a hermetically-sealed copper carrier tube, elastic cushioning fibers, a central member known as a king wire, and the fibers themselves.

Long-haul optical spans that crisscross oceans must be carefully designed to take into account geographical considerations, distance, and noise. Long spans must be amplified, but amplifiers amplify noise as well as the desired signal, and must therefore be carefully inserted so as to minimize the total noise injected by each of the amplifiers in the span. The noise is cumulative; each amplifier adds an additional noise component. To reduce the impact of this problem, submarine amplifiers tend to have less gain, allowing more amplifiers to be used in the span. The gain must be equalized across all wavelengths in systems where DWDM is employed.

Regeneration must also be done periodically to keep the signal clean, and in a typical installation, eight or so amplifiers are

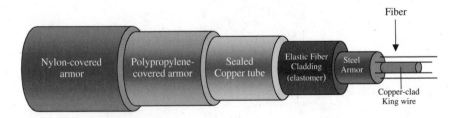

FIGURE 3-5 Submarine cable

installed in series before regeneration is performed. Because DWDM plays a major role in modern optical cable systems, dispersion and four-wave mixing effects, as well as other nonlinearities, must be carefully considered.

CABLE INSTALLATION OPTIONS

Cable installation techniques run the gamut from extremely simple to very complex. At the simple end of the installation spectrum, cable can be buried in a deep trench, or pushed into the ground by a high-pressure drill.

SUBMARINE INSTALLATIONS

Submarine cables are laid by special ships such as AT&T's *Global Sentinel,* shown in Figure 3-6. Once the cable has been laid by the ship and her crew, it must make landfall. Once the ship has pulled the cable to within a mile or two of shore, the cable is floated as shown in Figure 3-7. The floats bear the weight of the cable from the ship to the beach, where huge winches pull the end of the cable ashore with the help of terrestrial cable handlers (see Figure 3-8). Once ashore, the cable can be spliced into the network.

FIGURE 3-6 AT&T's *Global Sentinel* (Photo courtesy Todd Quam)

FIGURE 3-7   Floating cable (Photo courtesy Todd Quam)

FIGURE 3-8   Making landfall (Photo courtesy Todd Quam)

DUCTED CABLE

Optical cable can also be installed in ductwork by threading a fish line through the duct, attaching it to the cable that is to be inserted, and carefully pulling it into place. Cable that is to be pulled in this fashion is designed for the purpose, with special strength members that protect the fibers inside from undue stretching and bending. The ductwork usually has periodic access facilities along its route to enable for maintenance and splicing activities.

Plenum Cable

In much the same way that cable can be pulled through ductwork, it can also be pulled through plenum vaults, which are those spaces in a building that provide air transport. Because of the danger of fire and the resulting transport of smoke, cable that is to be used in plenum installations must be stringently rated to ensure that it does not give off smoke or hazardous fumes when heated or burned.

Aerial Cable

In environments where it is too costly to bury the cable or simply not feasible to dig, aerial installation is employed. Specially designed cable that has a stress-reducing strength member integrated into the cable can be strung between utility poles, attached to a separate messenger wire, or suspended by its own strength member, as with the figure-8 cable described earlier (refer to Figure 3-2). If a load-bearing messenger wire suspends the cable, it is securely lashed to the wire to prevent sagging. In all cases, optical cable designed for aerial installation must be able to handle the extremes of ice loading, wind shear, bird load (really!), deterioration from sunlight exposure, potential exposure to high voltage, and chewing rodents, especially squirrels.

Interior Low-Impact Installations

In areas where environmental concerns are at a minimum, such as in interior conduit, inside walls, above suspended ceilings, and in building cable risers, installation can be accomplished easily. A special ribbon cable that is impervious to foot traffic is designed to be installed beneath carpet.

COMMERCIAL FIBER PRODUCTS

A number of companies manufacture optical fiber cables in a remarkable array of flavors. These companies include Fitel, the joint venture company jointly owned by Lucent Technologies and The Furukawa Electric Company, Ltd.; Lucent Technologies itself; Corning; and Alcatel. These and others are discussed in the next section.

FITEL LUCENT TECHNOLOGIES

Fitel builds loose-tube optical cable assemblies for a wide variety of environments and applications, including high and low density cables, long and short-haul applications, as well as test and measurement equipment and fiber devices such as WDM systems, optical amplifiers, and specialized connectors. Their cables are as diverse as the applications for which they are used. These include single jacket construction for duct and lashed aerial use; light armored, for duct, lashed aerial and buried applications; armored, to offer protection from compressive forces and chewing rodents; double armored cable, to protect the fibers inside from burial in rocky terrain; and heavy armor, for extreme environment installations. They also include Mini-LXE Armored Sheath, for campus, LAN, CATV, and fiber ring installations; AccuTube™, designed for high-bandwidth applications that require the multiple fiber count that optical ribbons provide; OPTION1™, which offers low smoke and halogen rating for indoor installations; low moisture cable; Figure-8; and PowerGuide™, an all-dielectric self-supporting cable.

These cable assemblies can be built to transport multimode, standard single mode, TrueWave, and AllWave fiber, as required by the specific application. They can also be built with custom connectors.

LUCENT TECHNOLOGIES

Lucent is a leader in the optical fiber marketplace and boasts a wide array of products designed to address the entire spectrum of optical applications.

The TrueWave® fiber line is Lucent's flagship, offering a broad mix of products for many different environments. TrueWave® RS is an NZDF fiber designed to work in the third (C-Band) and fourth (L-Band) transmission windows. It provides uniform echromatic dispersion across both windows, low bending-induced loss at both 1550 and 1625 nm, and specifications for dispersion and attenuation in the fourth window. Its minimal dispersion slope characteristics enable it to reduce the impact of four-wave mixing while still maintaining long distance transmission capabilities.

TrueWave® XL Submarine Fiber is a negative dispersion, large effective area fiber designed for the rigors of long-haul submarine installations. It is a high-strength, low-loss fiber designed for use in the C-Band (1550 nm). Its specially selected refractive index provides a low, non-zero dispersion point to reduce the effects of four-wave mixing and enable the use of DWDM.

TrueWave® *Submarine Reduced Slope Fiber* (SRS) is also a negative dispersion fiber designed for effective use with EDFA-based systems. Its low dispersion permits the bandwidth of the system to be increased by allowing for more channels across the entire spectrum of available bandwidth. Because it is designed for submarine installations, it is rigorously pressure tested.

AllWave™ Fiber (see Figure 3-9) is a remarkably innovative product that opens a previously avoided transmission window. The 1400 nm band has always been avoided because of the presence of the 1385 nm hydroxyl absorption peak that occurs there. By manufacturing a hydroxyl-free fiber, Lucent increases the available window size by more than 50 percent and enables 10 Gbps operation in that band. This provides transport capability for low-cost CWDM systems, metro installations, and hybrid fiber-coax applications.

Corning®

Corning's name, of course, is associated with optical products, and their line of fiber is extensive. Corning manufactures both single-mode and multimode fiber in a variety of flavors for most networking applications.

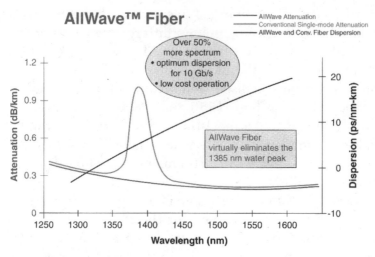

FIGURE 3-9   AllWave™ performance (Courtesy Lucent Technologies)

Corning's *Large Effective Area Fiber* (LEAF®) has received well-deserved praise in the last few years. It is designed for long-haul, high-bandwidth applications utilizing multichannel DWDM systems. As such it is targeted at long-distance and regional carriers, ILECs, CLECs, ISPs, cable television providers, and pan-continental service providers. LEAF® refers to the fact that the fiber enables the use of low light intensity, and therefore reduces four-wave mixing problems. It is designed to work effectively in the 1550 nm band where DWDM and multiple EDFAs are present. According to industry measurements, LEAF®'s effective area is more than 30 percent larger than other NZDSF products, offering lower attenuation and dispersion as a result.

LEAF® is also designed for submarine applications.

MetroCor™ is a negative dispersion fiber that operates in the 1550 nm band and is designed for metropolitan applications. Its negative dispersion characteristics enable rings to be lengthened without extensive dispersion compensation. It is designed to operate with lower cost distributed feedback lasers, thus helping to control the cost of network deployment.

SMF-28™ is Corning's single-mode fiber line designed for use in cost-effective environments that operate in the 1310 nm region. It can also be used in TDM and WDM systems that operate at 1550 nm. It offers low splice losses and can be ordered as loose tube, buffered tube, and ribbon configurations.

Corning also manufactures an extensive line of multimode fibers, including the InfiniCor® series as well as the Corning® 62.5/125 and 50/125 large core products.

ALCATEL OPTICS

Using their patented Advanced Plasma and Vapor Deposition fiber creation process, Alcatel manufactures and sells multimode fiber, single mode fiber, and numerous related products including DWDM equipment.

AN ASIDE: FREESPACE OPTICS

One transport alternative that has recently emerged is the concept of "fiberless-fiber." Also known as *freespace optics*, the concept revolves around the ability to transmit a coherent beam of laser light through the air from a source to a destination, eliminating the costly requirement to build a fiber infrastructure or to comply with FCC spectrum licensing requirements.

A number of companies are involved in this technology including AirFiber, TeraBeam, Infrared Communication Systems, AstroTerra, LSA Photonics, and LightPointe Communications. Bandwidth claims range from 155 Mbps (OC-3) to 622 Mbps (OC-12), over distances as far as 14 miles with availability ratings as high as 99.98 percent.

Freespace optics are somewhat limited compared to the fiber alternative: because the technology relies on line-of-sight transmission, distances are limited by 10 to 15-mile operational horizons due to the curvature of the earth, building construction, atmospheric conditions, and limitations on laser power due to safety restrictions. They are, however, enjoying success in the marketplace. CLECs seem to be the most significant customers, and the numbers are growing.

SUMMARY

Consider the following quotes from Stephen Garofalo of Metromedia fiber:

- Bits per second is history.
- The PSTN is dead.
- Unlimited bandwidth for a fixed price. That's our deal.
- Fifty times the bandwidth for less.
- We are spending more money than anyone else on thousands of fibers per route mile and eight conduits. We overprovision five times. Our vision of the world: waste bandwidth.
- First we took Manhattan. Now we're in Boston, Washington, Chicago, Dallas, Houston, Philadelphia, and San Francisco. Fourteen thousand buildings and 67 cities in the US and Europe.
- We turned the local loop into a virtual hard drive.
- You can deploy applications that were only a dream and you can't tell whether they are in your hard drive or in our fiber.
- We are the fibersphere.

Garofalo's colorful statements serve to underline what is happening in the optical networking world with regard to demand for high-speed, long-haul transport. Fiber and cable manufacturers are currently working around the clock to manufacture product, trying in vain to catch up to the demand that is driving the optical marketplace. Corollary manufacturers such as Huber & Suhner, JDS Uniphase/SDL, Avanex, Agere Systems, Texas Instruments, and a host of others are scrambling to manufacture connectors, pre-assembled cables, amplifiers, filters, tunable lasers, passive multiplexers, and other necessary components.

Of course, more is involved in this environment than transport. We must also take into account the need to multiply the available bandwidth as much as possible, the need to establish transport paths on demand, and the absolute need to manage the network. The first is done with DWDM; the second, with optical switching; and the third, with effective element and network management systems.

DENSE WAVELENGTH DIVISION MULTIPLEXING (DWDM)

When SONET was first introduced, the bandwidth that it made possible was unheard of. The early systems that operated at OC-3 levels (155.52 Mbps) provided volumes of bandwidth that were almost unimaginable. As the technology advanced to OC-12, OC-48, and beyond, the market followed Say's Law, creating demand for the ever more available volumes of bandwidth. There were limits, however; today, OC-48 (2.5 Gbps) is extremely popular, but OC-192 (10 Gbps) represents the practical upper limit of SONET's transmission capabilities given the limitations of existing time division multiplexing technology. The alternative is to simply multiply the channel count—and that's where WDM comes into play.

WDM is really nothing more than frequency division multiplexing, albeit at very high frequencies. The ITU has standardized a channel separation grid that centers around 193.1 THz, ranging from 191.1 THz to 196.5 THz. Channels on the grid are technically separated by 100 GHz, but many industry players today are using 50 GHz separation or less.

The majority of WDM systems operate in the C-Band (third window, 1550 nm), which allow for close placement of channels and the reliance on EDFAs to improve the signal strength. Older systems, which spaced the channels 200 GHz (1.6 nm) apart, were referred to simply as WDM systems; the newer systems are referred to as *dense* WDM systems because of their tighter channel spacing. Modern systems routinely pack 40-10 Gbps channels across a single fiber, for an aggregate bit rate of 400 Gbps.

How DWDM Works

As Figure 3-10 illustrates, a WDM system consists of multiple input lasers, an ingress multiplexer, a transport fiber, an egress multiplexer, and of course, receiving devices. If the system has eight channels such as the one shown in the diagram, it has eight lasers and eight receivers. The channels are separated by 100 GHz to avoid fiber nonlinearities, or perhaps closer if the system supports the 50 GHz spacing. Each channel, sometimes referred to as a lambda (λ, the Greek letter and universal symbol used to represent wavelength), is individually modulated, and ideally the signal strengths of the channels should be close to one another. Generally speaking this is not a problem, because in DWDM systems the channels are closely spaced and therefore do not experience significant attenuation variation from channel to channel.

A significant maintenance issue faces operators of DWDM-equipped networks. Consider a 16-channel DWDM system. This system has 16 lasers, one for each channel, which means that the service provider must maintain 16 spare lasers in case of a laser failure. The latest effort is underway in the deployment of tunable lasers, which allow a laser to be tuned to any output wavelength, thus reducing the number of spares that must be maintained.

External Cavity Tunable Lasers (ECTL) are the most common form of tunable light source in use today. Figure 3-11 shows a simplified diagram of an external cavity tunable laser.

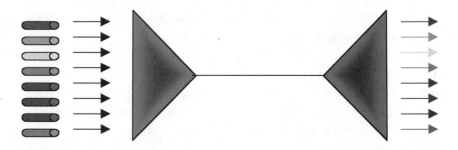

FIGURE 3-10   WDM Concepts

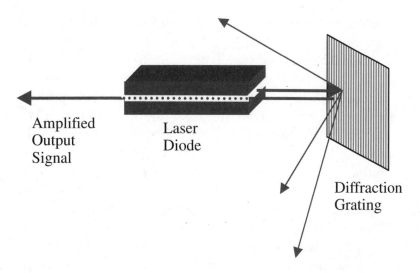

FIGURE 3-11   External cavity tunable laser

The laser comprises a laser diode, and a "rotatable" diffraction grating. The laser diode emits a range of wavelengths that must be tuned to a much narrower range to accommodate the stringent requirements of closely-spaced DWDM channels.

The diffraction grating is an etched array of fine lines, usually photolithographically "burned" into a segment of fiber by ultraviolet light, that can be "tuned" by rotating it so that it only allows certain wavelengths of light to be reflected directly back to the source—others, as shown in the diagram, are reflected at angles so that they do not enter the laser cavity for amplification and emission. You can simulate the effect of a diffraction grating by holding your index and middle fingers close together so that a thin space is between the first and second knuckles of both fingers. Look at a light source through this narrow gap, and by widening and narrowing the gap you will ultimately see the appearance of an array of fine black lines between your fingers. This interference pattern results from certain wavelengths of visible light being blocked by the narrow gap in the same way that a diffraction grating prevents certain wavelengths of light from entering the transmission system.

The grating receives the range of emitted wavelengths and reflects all but a narrow range of selected wavelengths away from the laser. The desired wavelengths are reflected back into the laser diode, where they are amplified and emitted from the output facet of the semiconductor. By rotating the Bragg filter, different wavelengths can be selected for transmission, thus creating a truly tunable laser.

So what do we find in a typical WDM system? A variety of components, including multiplexers, which combine multiple optical signals for transport across a single fiber; demultiplexers, which disassemble the aggregate signal so that each signal component can be delivered to the appropriate optical receiver (PIN or APD); active or passive switches or routers, which direct each signal component in a variety of directions; filters, which serve to provide wavelength selection; and finally, optical add-drop multiplexers, which give the service provider the ability to pick up and drop off individual wavelength components at intermediate locations throughout the network. Together, these components make up the heart of the typical high-bandwidth optical network. And why is DWDM so important? It's because of the cost differential that exists between a DWDM-enhanced network and a traditional network. To expand network capacity today by putting more fiber in the ground, on average, costs about $70K per mile. To add the same bandwidth using DWDM by changing out the end-point electronics costs roughly one-sixth that amount. Clearly a financial incentive goes with the WDM solution.

Many corporations in the industry manufacture DWDM multiplexers, including Lucent Technologies, Nortel Networks, Ciena, and many others. They will be examined in detail later in the book.

The next area of focus is switching and routing.

OPTICAL SWITCHING AND ROUTING

DWDM facilitates the transport of massive volumes of data from a source to a destination. Once the data arrives at the destination, however, it must be terminated and redirected to its final destination on a lambda-by-lambda basis. This is done with switching and routing technologies.

Switching versus Routing: What's the Difference?

A review of these two fundamental technologies is probably in order. The two terms are often used interchangeably, and in many cases a never-ending argument is underway about the differences between the two.

The answer lies in the OSI Protocol Model that we examined earlier in the book. Switching, which lies at Layer 2 (the Data Link Layer) of OSI is usually responsible for establishing connectivity within a single network. It is a relatively low-intelligence function and is therefore typically accomplished quite quickly.

Routing, on the other hand, is a Layer 3 (Network Layer) function. It operates at a higher, more complex level of functionality and is therefore more complex. Routing concerns itself with the movement of traffic between sub-networks and therefore complements the efforts of the switching layer. ATM, Frame Relay, LAN protocols, and the PSTN are switching protocols; IP is a routing protocol.

Switching in the Optical Domain

The principal form of optical switching is really nothing more than a sophisticated digital cross-connect. In the early days of data networking, dedicated facilities were created by manually patching the end points of a circuit at a patch panel, thus creating a complete four-wire circuit. Beginning in the 1980s, digital cross-connect devices such as AT&T's *Digital Access and Cross-Connect* (DACS) became common, replacing the time-consuming, expensive, and error prone manual process. The digital cross-connect is really a simple switch, designed to establish "long-term temporary" circuits quickly, accurately, and inexpensively.

Enter the world of optical networking. Traditional cross-connect systems worked fine in the optical domain, provided that no problem going through the O-E-O conversion process occurred. This, however, was one of the aspects of optical networking that network designers wanted to eradicate from their functional requirements. Thus was born the optical cross-connect switch.

The first of these to arrive on the scene was Lucent Technologies' LambdaRouter. Based on a switching technology called *Micro Electrical Mechanical System* (MEMS), the LambdaRouter was the world's first all-optical cross-connect device.

MEMS relies on micro-mirrors, which can be configured at various angles to ensure that an incoming lambda strikes one mirror, reflects off a fixed mirrored surface, strikes another movable mirror, and is then reflected out an egress fiber. The LambdaRouter is now commercially deployed and offers speed, a relatively small footprint, bit rate and protocol transparency, non-blocking architecture, and highly developed database management.

A schematic diagram of the MEMS technology is shown in Figure 3-12.

Nortel's OPTera Connect PX Connection Manager is another all-optical photonic cross-connect device. Nortel entered the optical cross-connect marketplace with its acquisition of Xros, and now offers a powerful product in the optical cross-connect space. The device eliminates all O-E-O conversion, is bit rate and protocol transparent, switches on a lambda-by-lambda basis, and is immensely scalable.

The mirror-based MEMS technology is the best-known wavelength switching technique, but other technological contenders have recently entered the marketplace and are showing great promise.

AGILENT TECHNOLOGIES

The Agilent Photonic Switching Platform includes a 32 x 32 port and a dual 16 x 32 port photonic switch, both of which include switch control and management electronics and a well-designed *application-programmer interface* (API). Based on a

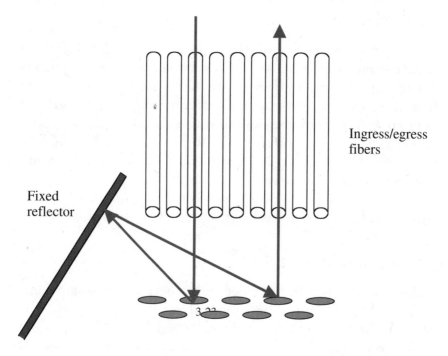

Ingress/egress
fibers

Fixed
reflector

FIGURE 3-12   MEMS operation

combination of the world-famous Agilent inkjet technology and
a relatively new technology called planar lightwave circuits, the
switch is capable of moving optical signals without the moving
parts found in MEMS-based devices.

The switch comprises an array of vertical and horizontal
waveguides that are permanently aligned. Optical signals are
transmitted across the horizontal waveguides from an input port
to an output port. When a switch command is received, a bubble is formed by applying heat at the intersection of the horizontal path along which the signal is traveling and the vertical
path that leads to the new output port.

Alcatel Optics and Agilent have joined forces to develop
switching elements based on this new technology. This relationship will lead to the development of optical cross-connect
switches, optical add-drop multiplexers, and optical-based protection switching equipment.

Other Switching Solutions

Manufacturer Gooch and Housego PLC has developed an acousto-optical switch that actually use sound waves to reliably deflect light from one fiber to another. Using a fused coupler that attaches two fibers to each other, specific wavelengths of light are forced from one fiber to the other. These devices have no moving parts and suffer very little loss compared to MEMS-based devices. They do have a downside, however: They are slow, and can be expensive.

Light Management Group is also working on an all-optical switch, and plans to release a commercial product shortly.

Routing in the Optical Domain

Today, routing is still an electrical function, although many modern terabit routers have optical network interfaces. The demand for their services is extremely high, and is being driven by the growing demand for the transport of rich media, a steady increase in the number of users, demands for greater mobility, and more reliance on IP-based services. According to RHK, communications traffic will grow 1700 percent or more by 2002 over 1998 numbers. High-speed routing will play a major role in the success of the network as it staggers to keep up with the demand.

The business case for terabit routing is easy to make. In a traditional network, a 16-channel DWDM device will interface with 16 OC-192-based add-drop multiplexers, which in turn will interface with a whole slew of lower-speed feeds. With the introduction of terabit routing, the add-drop multiplexers and lower-speed interfaces could be replaced with a smaller collection of terabit routers that interface directly to the network, as shown in Figure 3-13.

The players in the game are multiplying at dizzying speeds as well. The major players—Juniper, Cisco, Lucent, Avici, Sycamore, Nortel, and others—will soon face a host of new entrants backed by venture capital and looking to make a splash in what appears to be a long-term, lucrative market. In November

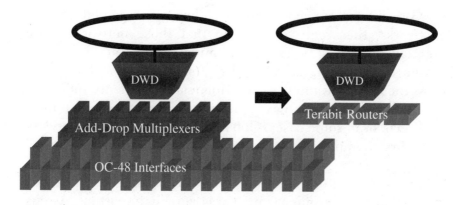

FIGURE 3-13   The case for terabit routing

2000, Pluris Inc. announced $100 million of fourth round fund-
ing that the company believes will be enough capital to ship its
Teraplex-based IP core product and carry them to the IPO point.
Others, such as Corvis, which bills itself as the pioneer of the all-
optical network, are covering the technological waterfront by
offering a complete line of optical networking products: an
Optical Network Gateway, optical amplifiers, optical add/drop
multiplexers, an optical switch, and a comprehensive network
management software package. Their goal is to provide all three
of the major infrastructure components: optical switching, optical routing, and network management.

NETWORK MANAGEMENT

Given the immense complexity of the evolving optical network
and the degree to which it contributes to the competitive positioning efforts of network providers, the first question that
must be asked about network management is "Why?" Why has
network management become one of the most important, yet in
many ways least successful efforts within the entire optical
domain?

When telephone companies first began to assemble complex networks, they realized early on that those networks represented enormous financial commitments. Their own senior management teams as well as the various regulatory bodies that governed them wanted to see financial statements indicating the success of the huge capital outlays required to build a national, or in some cases, global telecommunications network. The first component of network management therefore was financial management. Over time, however, particularly as competition edged into the telecommunications industry, it became clear that service management was equally important. As customers became more and more dependent upon the services they purchased via their service provider, and as those services became integral to their business operations and central components of their strategic plans, those same customers began to demand accountability from the service providers for guaranteed levels of service quality. *Service Level Agreements* (SLAs) that detail precisely what a customer can expect in the way of access, bandwidth, downtime, and compensation for failure to perform became the norm.

Today, network management is a key differentiator for many companies. Generally speaking, it addresses five key areas: fault management, configuration management, accounting management, performance management, and security management.

Fault management defines the set of processes required to detect, isolate, and identify the cause of a network failure.

Configuration management allows the service provider to ensure that the services delivered to a particular customer are specifically targeted at that customer's business issues. It also allows the service provider to bill for those services accurately.

Accounting management provides the mechanisms necessary to collect usage data that is then converted into billable elements.

Performance management ensures that each customer is receiving the level of service from the network commensurate with the commitments set forth in the SLA.

Finally, *security management* guarantees the integrity of the customer's data by ensuring that appropriate steps are taken to safeguard the network and its hosts from date interception and unauthorized intrusion.

THE CHALLENGES OF NETWORK MANAGEMENT

Because of the diverse audiences that require network performance information and the importance of SLAs, the data collected by network management systems must be malleable so that it can be formatted for different sets of corporate eyes. For the purposes of monitoring performance relative to service level agreements, customers require information that details the performance of the network relative to the requirements of their applications. For network operations personnel, reports must be generated that detail network performance to ensure that the network is meeting the requirements of the SLAs that exist between the service provider and the customer. Finally, for the needs of sales and marketing organizations, reports must be available that enable them to properly represent the company's capabilities to customers, and to enable them to anticipate requirements for network augmentation and growth.

One of the greatest challenges that face network management systems is the diversity of players in the game. In their zeal to ensure manageability, most device manufacturers included element management systems as an integral part of each device (element) they designed and marketed. For example, a modem pool, a T-1 multiplexer, a router, and a local switch, all from different vendors, have element management databases that allow a higher-level process to request and collect information about the health of each element in the network. The problem with this approach is that these vendor-provided systems are largely proprietary, and therefore do not interoperate with other element management systems. The result is that a network manager is required to simultaneously monitor outputs from a variety of element managers, and when problems occur, mentally combine the collective outputs, analyze them, and rapidly (and correctly!) discern a corrective action plan. Clearly, a better way to do business exists.

Consider the drawing shown in Figure 3-14. Here we see a

three-tiered system with the managed elements at the bottom, the element management systems in the middle, and the actual network management system at the top. Functionally, the three are different, but highly related. The elements are the devices that are to be managed; they inherently collect management information in dedicated, proprietary databases called *Management Information Bases* (MIBs). These MIBs are periodically interrogated by the *element management systems* (EMS), which collect and package the information for eventual delivery to the actual network management system. The network management system receives all of the data, massages it, analyzes it, and creates information that a network manager can use to make quick and accurate decisions to preserve network integrity and ensure compliance with SLAs.

The biggest problem with network management is that the elements in a network, which may well be manufactured by different vendors, deliver information to the EMS in proprietary form. For purposes of information integration, it is critical that they deliver their data to the uppermost tier in a form that it can

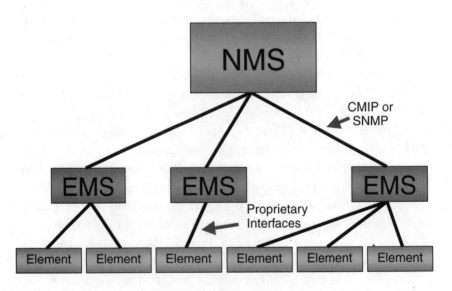

FIGURE 3-14   The network management hierarchy

easily understand. In fact, *all* of the EMS must deliver their reports in a common form.

Both ISO and the IETF created solutions to this problem of vendor interoperability. ISO released a protocol known as the *Common Management Information Protocol* (CMIP), although the TCP/IP-oriented IETF released the far more successful *Simple Network Management Protocol* (SNMP). SNMP has gone through numerous iterations as it has matured. Today it is the most widely accepted "canonical form" protocol for the delivery of network management data to a higher-level management system for processing.

Other standards have entered the management pantheon in recent years. For some time now, the *Telecommunications Management Network* (TMN), an offshoot of ISO's CMIP, has been considered a good model for network management. As the network profile has changed, however, and with the steady migration to IP and a renewed focus on service rather than technology, the standard TMN philosophy has begun to appear somewhat tarnished.

Originally designed by the ITU-T, TMN is built around the OSI Model and its attendant standards, which include the *Common Management Information Protocol* (CMIP) and the *Guidelines for the Development of Managed Objects* (GDMO).

TMN employs a hierarchical model that comprises a *network element layer,* an *element management layer,* a *network management layer,* a *service management layer,* and a *business management layer.* Each has a specific set of responsibilities closely related to those of the layers that surround it.

The *network element layer* defines each manageable element in the network on a device-by-device basis. Thus, the manageable characteristics of each device in the network are defined at this functional layer.

The *element management layer* manages the characteristics of the elements defined by the network element layer. Information found here includes activity log data for each element. This layer houses the actual element management systems responsible for the management of each device or set of devices in the network.

The *network management layer* has the ability to monitor the entire network based upon information provided by the element management layer.

The *services management layer* responds to information provided by the network management layer to deliver such service functions as accounting, provisioning, fault management, configuration, and security services.

Finally, the *business management layer* manages the applications and processes that provide strategic business planning and tracking of customer interaction vehicles such as service level agreements.

OSI, although highly capable, has long been considered less efficient than IETF management standards, and in 1991, market forces began to effect a shift. That year, the Object Management Group was founded by a number of computer companies including Sun, Hewlett-Packard, and 3Com. Together they introduced the *Common Object Request Broker Architecture* (CORBA). CORBA is designed to be vendor-independent and built around object-oriented applications. It enables disparate applications to communicate with each other, regardless of physical location or vendor. And although CORBA did not achieve immediate success, it is now widely accepted, resulting in significant CORBA-based development efforts among network management system vendors. Although this may seem to fly in the face of the OSI-centric TMN architecture, it really doesn't. TMN is more of a philosophical approach to network management, and does not specify technological implementation requirements. Thus, a conversion from CMIP to SNMP, or the implementation of CORBA, does not affect the overall goal of the TMN.

NETWORK MANAGEMENT IN THE REAL WORLD

For all its importance in the network, most manufacturers of optical telecommunications equipment do not have a good story to tell. Plagued by the need to roll out hard products at high speed to counter the efforts of the competition, many companies have failed to give network management its due. Most companies offer good element management systems; few have created integrators that have the ability to effectively integrate the input from diverse EMS and create meaningful output for network managers of the human variety. Some have management systems that only manage certain types of hardware, or certain regions of the network; but in a world where end-to-end management of complete solutions is the most desirable thing to offer customers, this "hole in the product line" is a serious liability. Some of the smaller companies such as Sycamore, Amber Networks, and Astral Point have made network management the centerpoint of their product offerings.

Lucent's WaveWrapper technology is a step in the right direction with regard to management of the optical network. In the telco domain, SONET offers significant management capability. However, given advances in modern networking, it is now viewed as being somewhat overhead intensive, difficult to scale, and inflexible. A considerable effort is afoot to move some of the management functions down into the "pure" optical layer, thus eventually eliminating the need for SONET's overhead-heavy management bytes and allowing network providers the ability to manage at the lambda level. WaveWrapper does exactly that. It "wraps" the optical transmission stream within a digital "wrapper" that carries information that can be used to monitor individual wavelengths within a DWDM-encoded optical stream and perform *Forward Error Correction* (FEC) and channel performance monitoring as required, without the need for retransmission. So successful has been WaveWrapper that it is being considered the prototype for an ITU standard to define generic digital wrapper techniques.

Other efforts are underway, as well. The "Optical Network Navigator" concept is designed to take advantage of existing, in-place SONET and ATM EMS. Its ultimate goal is to provide dynamic wavelength allocation, under which virtual circuits provisioned in ATM will flow through the system into the optical layer to bring about dynamic trunking management at the optical level. The bottom line is this: network management is typically the last thing funded, the last thing created, the last thing trained on, and the last thing provisioned. Yet it represents the point at which the network most effectively touches the customer and the customer's requirements. Network equipment manufacturers would be wise to move it to the top of the importance hierarchy. They would also be well served to recognize and act upon the differences between *element management systems* and *network management systems*. Most companies have the first; few have the latter.

PUTTING IT ALL TOGETHER

Let us review once again the service requirements that are being placed on the network today:

- The need to create routes on demand between individual users as well as between disparate work groups, in response to the market shying away from dedicated, costly facilities.
- Guaranteed interoperability between disparate protocols.
- Universal, seamless connectivity between far-flung corporate locations.
- Optimum utilization of network bandwidth through the appropriate use of intelligent prioritization and routing techniques.
- Traffic aggregation for wide area transport to ensure efficient use of network bandwidth.
- Granular quality of service control through effective policy and queue management techniques.
- Growing deployment of high-speed access technologies such as DSL, cable modems, wireless local loop, and satellite connectivity.

These requirements point clearly to a limited set of critical observations about the nature of the evolving optical network. First, it must be able to offer seamless, high-bandwidth, protocol-independent transport among all four major regions of the global network: enterprise, access, terrestrial transport, and

submarine transport. It must provide highly differentiable quality of service levels, and must support interoperability with a
wide variety of access and switching/routing technologies.
These requirements culminate in the combination of the technological piece parts described in this section: diverse optical
cable solutions; the use of DWDM where appropriate as a technique for effective bandwidth multiplication; optical switching,
routing, and add-drop as a way to move customer traffic without
the need for optical to electrical conversions. They also include
optical amplification, a way to reduce the cost of cable deployment; and network management, the functional fabric that ties
the network together and bridges the gap between the provider
of services based upon optical technologies and the client that
needs appropriately targeted services.

As we will see in the next section, a hierarchy of needs in
telecommunications is analogous to Maslow's hierarchy of
human motivation. Maslow observed quite correctly that
humans are motivated by a desire to achieve higher-level
accomplishments, but only after lower level requirements
such as safety and sustenance have been satisfied. We begin
the next section with a discussion of "the optical hierarchy of
motivation."

PART FOUR

SOLUTIONS AND APPLICATIONS

THE OPTICAL HIERARCHY OF MOTIVATION

In the early 1960s, Abraham Maslow, an American psychologist, proposed his now-famous hierarchy of basic human needs.

Maslow defined a clear six-level hierarchy of motives that determines the directions of human behavior. In his hierarchy, he ranked the basic human needs as follows: physiological demands; security and safety concerns; love and feelings of belonging; competence, prestige, and esteem; self-fulfillment; and finally, curiosity and the need to understand the surrounding world. Maslow concluded that the basic requirements had to be fulfilled before the higher level, more cognitive demands could be fulfilled.

Interestingly, a similarly structured "hierarchy of needs" in the domain of optical networking exists, particularly as it relates to the activities of network designers, service providers, and manufacturers of components and network devices. This hierarchy, shown in Figure 4-1, illustrates seven-layers: features; functions; benefits; applications; services; solutions; and value. All of these words appear in various marketing, sales, and technical documentation; and all are important. However, the manner in which they and the qualities that they each represent are used, are diverse and critically important.

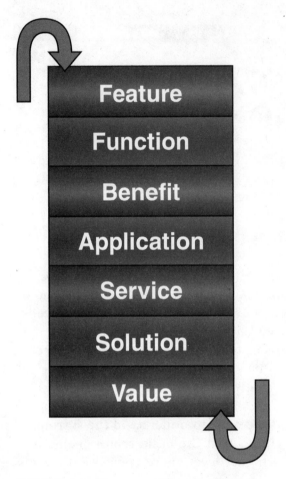

FIGURE 4-1   The network hierarchy of needs

From the dictionary, we derive the following definitions of each word:

Feature: A prominent or distinctive aspect, quality, or characteristic.

Function: The action for which (something) is particularly fitted or employed.

Benefit: Something that promotes or enhances well-being—an advantage.

Application: The act of putting something to a special use or purpose.

Service: An act of assistance or benefit to another or others.

Solution: The method or process of solving a problem.

Value: Worth in usefulness or importance to the possessor—utility or merit.

Two ways are used to approach the use of this network services hierarchy. One way is from the top down, starting with features and passing through the various layers of the model. The other is to start with value at the bottom and work upward. Both are valid, but must be utilized appropriately. Let's consider the functional differences between the seven layers.

Features, functions, and benefits are characteristics that define the technical capabilities and defining parameters of a device or service. "A prominent or distinctive aspect, quality, or characteristic," the definition of a feature, clearly speaks to such things as physical footprint (space a device occupies in a central office), amount of heat it generates while in operation, amount of electricity it consumes, backplane capacity, and component redundancy. Function, "the action for which (something) is particularly fitted or employed," describes the technical inner workings of a device or software module that result in some sort of operational value for the purchaser of the product. A benefit, "something that promotes or enhances well-being—an advantage," provides precisely that: an advantage that the product conveys to the customer, although typically interpreted as a technical rather than a market advantage, such as the ability to perform a hot swap of a component.

Moving down the stack, we come to application, "The act of putting something to a special use or purpose." An application refers to the manner in which a product is actually used—usually the reason that the customer purchased the item in the first place and the first occurrence of possible value to the consumer. Service, "an act of assistance or benefit to another or others," refers to the process of converting what is usually a generic application set into a more focused, almost customized treatment to address a specific set of needs for a client. Solutions, defined as "methods or processes used to resolve a problem," carry the specificity of services to the next level, addressing the customer's specific business requirements and often presenting a product directed as much at the customer's customer as at the actual customer.

Finally, we come to value, something that is useful or important to the possessor and offers utility or merit. Value is the most critical of the seven elements, because it is universal and timeless. Value, custom-defined in the mind of each customer, is a personal, specific, and difficult-to-quantify essence unique to every client. The other six—feature, function, benefit, application, service, and solution—are relatively static and time-dependent. In other words, an application that is timely, useful, and effective today may not be six months from now. A solution that resolves today's vexing customer problem may not this time next year. Value, however, is a constant, and has little to do with technology. It has everything to do, however, with a discrete knowledge and understanding of what makes the customer tick.

A number of other analogies lend themselves to this discussion. Consider for a moment the OSI Reference Model (see Figure 4-2). Its seven protocol layers cover the waterfront from physical transmission of bits to the lyrical interpretation of the meaning of specific data structures at the Application Layer. However, another hierarchy is at work here that is often overlooked—the hierarchy of customer proximity.

The lower three layers of the OSI Model—Physical, Data Link, and Network—have the following characteristics. They are based on clearly defined, widely accepted international standards. Relatively little room exists for interpreting their very clearly-stated intentions. They rarely change, are deeply embedded in the core of the network itself, and with few exceptions, rarely touch the customer. Finally, they represent the current operational domain of the typical service provider. Transport,

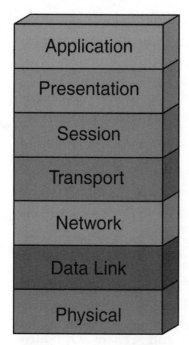

FIGURE 4-2   The OSI Reference Model

switching, and routing are the responsibilities of the telephone companies, cable companies, and their various transport cousins.

The upper layers, on the other hand, have varied responsibilities and are characteristically quite different from the lower layers. They are open to broad interpretation because they are close to the customer and must therefore be able to accommodate the diverse requirements of diverse application types. Although based on standards, the standards are fluid and are constantly being augmented or modified to meet the changing demands of the clients they serve. They are found scattered around the periphery of the network, because that's where the clients are. Finally, they represent the operational domain of content providers, service providers (in the truest sense of the term), *application service providers* (ASPs), and other companies that rely on the underlying network to transport their traffic—highly customized for each customer. Clearly, then, a functional separation occurs between the bottom and top of the OSI food chain. At the bottom, the network reigns supreme. At the top, the customer is emperor of all that he or she surveys.

In the days when bandwidth held sway and was a tremendous money generator, the lower layers represented a cash cow ripe for milking. Today, however, with the perceived glut of bandwidth that is now available thanks to the optical network providers, the price of bandwidth is plummeting to near-zero, a frightening reality for those companies that have traditionally made their fortunes through the sale of bits-per-second. Today, the big money lies at the top of the stack, closer to the customer, a place where unique, specially-designed network products can be positioned on a customer-by-customer basis.

The traditional service providers—the ILECs, CLECs, and IXCs—are scrambling to establish a toehold that will enable them to climb the stack, to move out of the primordial network ooze into the lofty heights of content and services. Of course, they are in an enviable position if they play their cards correctly. They touch the customers with their networks, and as my friend and colleague Dave Hill observes, "When you've got them by the access lines, their hearts and minds will follow." The combination of network provisioning and content is unbeatable, and is fast becoming a major focus for converged providers. The ability to provision an end-to-end network as well as deliver application content is a powerful combination. Thus, the service providers' interest in moving up the food chain is a valid one.

They do have a challenge before them, however, illustrated in Figure 4-3. The illustration shows two triangles. One is labeled "Knowledge," the other, "$." The Knowledge pyramid has a broad base, while the $ pyramid has a broad top. Above the pyramids we find the Upper Layers, below, the Lower Layers. The significance of this graphic is that the traditional service providers enjoy a broad base of knowledge about the lower three layers, given the fact that they have been providing them for 125 years. Furthermore, it costs them very little to operate there, because of their massive embedded base.

Upper Layers

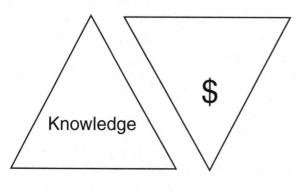

Lower Layers

FIGURE 4-3   The capability hierarchy

On the other hand, they have precious little knowledge about the upper layers, and the accumulation of that knowledge will be an extremely expensive undertaking, albeit a necessary one. The process of adding capability within the higher layer services to their collection of existing skills is a necessary next step for long-term viability.

The content providers and ASPs have the opposite problem. Their knowledge of upper layer services is quite rich and well developed, but they have little if any network capability. The capital investment required to build a network of their own would be prohibitive, which explains why so many of them are forming alliances or ownership arrangements with network providers. Consider Qwest, for example. As a "Bandwidth Baron," they have a globally well deployed optical network, as well as content capability through CyberSolutions, their joint venture with KPMG. Their acquisition of USWest enables them to cover the access, transport, and services waterfront—an enviable combination of capabilities.

Yet another example of this evolution exists: the ongoing inversion of the network. As intelligence, capability, and bandwidth move inexorably away from the core toward the edge of the network, the margins of the network cloud expand as they strive to touch the customer. A relatively small collection of centralized functions, delivered by core switches and shared among a large collection of users, tends to treat the customer as a commodity and makes the assumption that their collective service requirements will be largely the same. The process of migrating capability closer to the customer dashes this philosophy on the rocks and enables services to be customized to a highly granular degree. David Isenberg, author of "The Rise of the Stupid Network," predicted this evolution years ago in his seminal work. In it, he observes that

A new network "philosophy and architecture" is replacing the vision of an Intelligent Network. The vision is one in which the public communications network would be engineered for "always-on" use, not intermittence and scarcity. It would be engineered for intelligence at the end-user's device, not in the network. And the network would be engineered simply to "Deliver the Bits, Stupid," not for fancy network routing or "smart" number translation.

These customized services take advantage of the redesigned network and its capabilities, and redefine the relationship between customer, provider, and services.

So what does all this mean? It means that the winners in this game to acquire and keep customers will be those companies that understand the importance of providing highly targeted value based upon a discrete understanding of the drivers behind every customer. Technology, expressed as features, functions, benefits, and so on, is still critically important. However, the people who care about that level of technical detail are not generally the people who make buying decisions or write checks for product purchases. Value takes on many different forms, and must be expressed appropriately to each audience. Technology bells and whistles certainly matter—I would never suggest otherwise—but much more to the capability equation exists. The characteristics that represent value to a technician responsible for installing, operating, and maintaining an optical device are different than those that provide value to a network designer, a beneficiary of the services that the device delivers, or the end user who must make a buying decision about a highly capital-intensive acquisition. Thus, the manner in which each product is represented to the client must be customized to an appropriate degree if the sale is to be successful.

This evolution, illustrated by the "separation of duties" in the OSI Model, the continuum of capabilities of the value equation, and the ever-expanding network and its migration of capability toward the customer, represent the inexorable need for customization. A need for knowledge about the customer and the ability to interpret the knowledge and respond with solutions provide undeniable, targeted value to each and every client's individual needs.

An electrical concept called "skin effect" provides a powerful analogy here. It is a well-known fact that current does not flow in the core of a conductor. Instead, it flows on the surface, because of the reduced resistance owing to the far greater surface area that is available there. This is why stranded wire is often considered to be a greater conductor than solid. The multiple strands of wire provide enormously greater surface area than a single solid conductor.

The analogy works well here: Far more market opportunity "surface area" exists on the surface of the cloud than within. Although the core marketplace continues to be important, the opportunities, the growth, and the money, lie at the edge.

The industry, has segmented itself into a hierarchy of functional players that work together to satisfy the changing demands of a remarkably diverse client base. A food chain is at work through which the various players combine their efforts in a form of value-added service to satisfy customer demands. The collective efforts of the various manufacturing layers and those of the service providers result in the foundation of what is required to create value in the eyes of the customer. That value derives from stable, reliable hardware and software, functional applications, and a clear focus on the customer's value chain. In the next section, we examine the players in this game and the roles that they play.

PLAYERS IN THE NETWORK GAME

The optical network is made up of four primary market segments: the component manufacturers who design and create ICs and optoelectronic devices, sometimes referred to as device manufacturers; the device manufacturers who build routers, switches, multiplexers, and other optoelectronic network devices, sometimes called system manufacturers; the big fiber houses; and the service providers, who design and build networks using the aforementioned devices. At the absolute top of the pyramid, of course, are the end users, the clients, and customers that benefit from the efforts of all the subtending layers (see Figure 4-4).

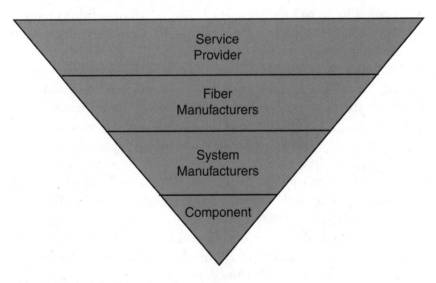

FIGURE 4-4 The Optical market segments

COMPONENT MANUFACTURERS

Optical technology is considered by many to be expensive and most segments, led by the efforts of the component manufacturers, are focusing renewed efforts on the challenge to reduce the cost of optoelectronics. The cost question is largely erroneous, particularly when optoelectronics and copper electronics are compared on the basis of raw bit transport. Nevertheless, a strong perception of inordinately high cost must be managed and ultimately overcome.

This segment of the optical market is dominated by a number of players including JDS Uniphase/SDL Inc., Lucent, Agere Systems, Nortel Networks, Corning, and a number of others. They are driven by the demands of the device manufacturers who integrate their components into finished product. All of these companies understand the Law of Primacy—the first one to market wins—and want to satisfy the burgeoning demands of customers. This requires that component players create integrated solutions that bring together the capabilities of transmitters, receivers, modulators, multiplexers, and amplifiers in a single package, a challenge that has been met with remarkable results. As the metro and long-haul markets collide, the need for sophisticated routing and network management functions has become more critical. As traffic requirements have increased, the demand for denser channel capabilities in DWDM systems has climbed in lockstep. All of these make for a remarkably complex environment. The companies that will win this part of the game are the companies that recognize the importance of converged capabilities and that have it all: research and development, strong marketing, and knowledge management organizations. They also have in-house manufacturing (or a closely-held manufacturing partner relationship), and a strong sales organization with the ability to anticipate the market and drive management to respond quickly. A recent spate of consolidations has affected this segment of the market. Most recently, JDS Uniphase acquired SDL for $41 billion as a way to round out their own product offering. Stay tuned—this is a lucrative market and more will follow.

According to Ryan, Hankin, and Kent (RHK), the market for DWDM components is set to grow from $5 billion in 2000 to more than $24 billion by 2004. Already, the market for amplifiers, filters, and wavelength splitters has grown more than 100 percent, and according to RHK, no single component segment will grow less than 80 percent.

For the major component manufacturers, this is good news. It is also timely for end-users, who more than anyone else will benefit as more and more optical components enter the market. System integrators at the same time will be able to devise services-oriented optical systems, which will increase bandwidth and lower the overall cost for service providers.

The forecasts for component demand are being driven primarily by increasing demand for DWDM devices and by the unquenched thirst for bandwidth, such as 10 GBps and beyond. Faster components lead to greater complexity, such as more capable amplifiers with better dispersion management capabilities.

A strong demand is growing for both active and passive optical components. According to RHK, Raman gain modules, 40G bit/sec active components, and tunable lasers will soon be commonly used in DWDM equipment, and by 2003, will make up $4 billion of the entire DWDM components market.

Some components will exhibit greater growth curves than others. According to RHK, tunable lasers, for example, will grow from a $280 million market in 2001 (up from $2 million in 2000) to $960 million by 2004.

KEY PRODUCT AREAS

The primary focal points within the components domain span a remarkable range of functionality. They include tunable lasers, Bragg filters, Faraday rotators, erbium-doped fiber amplifiers, transponders, and "systems on a chip," that is, the ability to consolidate multiple functional devices on a single chipset— such as an entire add-drop multiplexer on a piece of silicon.

Tunable lasers have become points of focus because of the degree to which they can reduce the cost of provisioning spare lasers for DWDM installations. For example, if an 80-channel DWDM system is installed, the service provider must stock a minimum of 80 single-channel lasers to guard against the possibility of channel loss. If, however, the company uses tunable lasers in their installation, they can dramatically reduce the total number of required spares, because the lasers can be dynamically tuned to the appropriate wavelength.

Bragg filters and Faraday rotators are in great demand because of their ability to selectively and passively route specific wavelengths of light. Bragg filters, described in the prior section, play a major role in tunable laser technology. Faraday rotators are slightly different. Instead of selecting wavelengths by creating an interference pattern (the Bragg concept), they use a combination of polarization effects and magnetic field manipulation to selectively remove and reroute individual wavelengths to one or more output fibers.

An area where this technology may play a critical role is *Passive Optical Networking* (PON), a relatively new arrival on the transport scene. *Fiber to the Home* (FTTH) or small business has been talked about for a very long time, but to date the cost of installing fiber and active optoelectronics devices has been prohibitive. For years, the search has been on to find an inexpensive alternative to active optoelectronics, and the search may be coming to a close with PON. For distributive networks, such as cable television systems, PON may provide a cost-effective solution by using a single, high-power laser source to provide content to many endpoints through some form of splitter or passive coupler. Lower speed return channels could be accomplished through the use of less expensive, lower power lasers.

Alternatively, each home or office could have a dedicated modulator designed to enable only the subscribed services to be delivered to the premises equipment, similar to the electronic notch filters used today in cable systems.

Erbium-Doped Fiber Amplifiers (EDFA) and Raman amplification are also important today, particularly for the growing demands of ultra-long haul transport systems that suffer from distance-dependent signal degradation. Demand for EDFAs in particular is extremely high, while Raman amplification, still a relatively nascent technology, is just beginning to grow in importance.

Transponders, the next area of significant interest in the component arena, represent a form of technological convergence that is centrally important, particularly when space considerations are key. Transponders are integrated silicon devices that include transmitters, receivers, multiplexers, and demultiplexers. In the switched world they serve as a replacement for traditional line cards, including as they do the functional equivalent of a transmitter, a receiver, and a multiplexer. In many cases they interface directly with onboard microprocessors for greater speed and functionality, and are expected to provide adequate capability in the future to completely replace an entire shelf assembly, such as a SONET or SDH add-drop multiplexer. OC-48 (2.5 Gbps) and OC-192 (10 Gbps) transponders are widely available, and OC-768 (40 Gbps) are expected to appear in the product lineup of many manufacturers in the near future.

THE PLAYERS

At the component level, seemingly countless players exist in the optical networking game. One of the largest and most diverse is Agere Systems (formerly Lucent Microelectronics), which manufactures a broad line of devices including fiber amplifiers, continuous wave and electroabsorptive modulated tunable laser transmitters, lithium niobate modulators, receivers (with and without clock recovery), all on a small form-factor. They also manufacture high-speed (40 Gbps) transceivers, transponders, 10 Gbps integrated circuit arrays for optical cross-connects, and a variety of other devices designed for specific applications. Others, such as JDS Uniphase, produce an equally diverse range of products, thanks in part to well-planned acquisitions designed to add capability to the corporate product line. Lucent, Cisco, and Nortel have collectively spent more that $25 billion on optical technology companies since the beginning of 1999, and others will follow. JDS recently purchased MEMS manufacturer Cronos Integrated Microsystems for $750 million, a strategic move designed to move the company into the burgeoning all-optical switching domain. Corning, meanwhile, recently purchased Pirelli's components division for $3.6 billion, while at roughly the same time announcing a $20 million expansion of their Garden Grove, California-based Advanced Photonics Facility. The plant manufactures LCD components used in amplifiers and DWDM systems. Currently, the plant manufactures an LCD-based, 80-channel system called the Wavelength Selective Switch, as well as the Dynamic Spectral Equalizer, which manages power in DWDM equipment. They have also invested heavily in MEMS technology, which they plan to deploy in optical cross-connect systems. Their LCD products, meanwhile, will be used for channel-to-channel switching in high channel count DWDM systems. Competing with Corning in the LCD business is Chorum, as well as Cypress Semi, which manufactures MEMS components through its recently acquired Silicon Light Machines subsidiary.

A number of other players have made impressive names for themselves as well. Switzerland-based Huber & Suhner manufactures a broad range of high-quality optical fiber assemblies. Infineon, formerly Siemens Semiconductors, produces optical fiber components. LightLogic designs and manufactures 10G transponders for SONET and Gigabit Ethernet applications. Kymata Ltd., located in Scotland, manufactures optoelectronic devices for DWDM applications, as does APA Optics. Kymata recently announced a multi-million dollar deal with Marconi in which Kymata will sell arrayed waveguide Bragg gratings, variable optical attenuators, optical wavelength power monitors, and thermo-optic switches. They will be used in Marconi's SmartPhotoniX DWDM systems for long-haul, regional, and metropolitan applications.

Other companies include Broadcom, a leading manufacturer of optical networking products including Gigabit Ethernet chipsets and other devices for local, regional, and metropolitan area networks. Nanovation Technologies, another key player, produces the Wide Band Nanoshutter™ Optical Switch and the Single Mode Wide Band Planar Optical Splitter, in addition to a variety of EDFAs, add-drop filters, attenuators, couplers, and DWDM multiplexers. Southampton Photonics, another player, manufactures Bragg gratings, amplifiers, and DFB laser arrays for DWDM installations. Novalux, a relatively new entrant, has made its name manufacturing laser modules, particularly its innovative *Novalux Extended Cavity Surface Emitting Laser* (NECSEL) that offers extremely narrow emissions at a very high tuning rate. Qusion Technologies, another player, manufactures indium phosphide-based chipsets, including a 40 Gbps optical modulator and a 1×2 switch module.

Like any industry, a few companies push the edges of the design and development envelope. Optune Technologies, a subsidiary of StockerYale, recently announced its intent to manufacture solid-state, tunable optical filters. They claim that their tunable filters will be better than existing designs on several levels. Some tunable filters are tuned by either heating or stretching the Bragg-treated fiber, which is slow and can ultimately lead to failure. Others are tuned mechanically, which makes them slow and expensive. The key to Optune's technology has not yet been announced, but the market is interested.

KEY ISSUES FOR COMPONENT MANUFACTURERS

Component manufacturers in the optical networking world face one overarching challenge—success. The optical transport was worth $12.3 billion in 1999, and shows no signs of slowing down any time soon. According to California consultancy RHK, the optical market as a whole, which includes SONET equipment, DWDM, and digital cross connect systems, grew 56 percent in 1999. These numbers will surprise those who believe that SONET is in its declining years. According to Cahner's In-Stat Group, the SONET market will increase from $11.89 billion in 1999 to $31.26 billion in 2004, because of its embedded importance to interexchange carriers, local exchange carriers, and Internet service providers.

Optical networking truly has reached a fever pitch, and manufacturers at all levels are scrambling to satisfy the growing demand for components, systems, and fiber. Globally, major projects are on indefinite hold while they wait for delivery of critical optical networking components. At the time of this writing, it is not uncommon for companies ordering certain products to be given delivery dates more than a year out, while manufacturing facilities toil around the clock throughout the year.

The other problem that component manufacturers face is differentiation. Optical devices, once available from very few providers, have now become commodities of a sort because of the number of companies manufacturing them. As a result, they are often backed into a corner and forced to become marketplace leaders by selling for the lowest possible price. This is not a desirable market to play in, and the result is that many of these companies are trying to move up the proverbial value chain by creating integrated systems on a chip that satisfy multiple customer requirements. These innovations have proven to be extremely successful in the marketplace, but require that companies shift their focus away from the traditional component market and focus instead (or additionally) on the more complex market that resides at the next level of the services food chain.

The final issue that component manufacturers face is the ongoing drive to further miniaturize the components that they sell. Space today is at a premium, and anything that can be done to reduce the footprint required for device installation is well-received by service providers as a way to reduce their costs.

SYSTEM MANUFACTURERS

Dependent as they are on the component manufacturers, the system manufacturers face a parallel challenge. They too must make product available to the service providers, but are hampered by high demand. They reside in a ferociously competitive segment of the optical networking market, and must take extraordinary and often highly innovative steps to retain customers in the face of equally creative actions by their competitors.

KEY PRODUCT AREAS
The device or system manufacturers are largely engaged in the design of high-speed switches and routers, including DWDM, SDH, and SONET products such as DWDM multiplexers and

add-drop and terminal multiplexers; terabit routers; and wire-speed switches. This field will inevitably go through its own wave of consolidation as the feeding frenzy continues. Furthermore, advances in all-optical switching have set this segment on fire. Innovations have poured out of this arena, including the *Micro Electrical Mechanical System* (MEMS), a chip-based array of microscopic mirrors that can be gimbaled under the command of a router to redirect optical inputs from one fiber to another. Agilent's bubblejet-based optical switch is another, which uses tiny bubbles to change the path of an optical input; and most recently, Gooch and Housego and Light Management Group's acousto-optical switching systems, which use sound waves to redirect optical inputs.

The players in this segment include such notables as Lucent, Cisco, Nortel, Ciena, Marconi, Sycamore, Alcatel, Juniper, and many others. All have positioned themselves to satisfy one or more segments of the optical networking environment, and to various degrees have been successful. The remarkable thing is the degree to which acquisitions have broadened the capabilities of these companies. Nortel, Lucent, and Cisco alone have been engaged in an optical feeding frenzy, as evidenced by the companies they now own. Between December 1998 and May 2000, Nortel acquired Cambrian Systems, Qtera, Xros, CoreTek, and Photonic Technologies. Lucent, meanwhile, acquired Ignitus and Chromatis, and Cisco picked up SkyStone Systems, PipeLinks, Monterey Networks, Cerent, Internet Engineering Group, Pirelli Optical Systems, Growth Networks, and Qeyton Systems.

The largest players in the game are Lucent, Nortel, Alcatel, and Cisco, with Lucent and Nortel holding more than 50 percent of the market. All sport impressive product offerings that will be covered below.

AGILENT TECHNOLOGIES

Agilent came booming into the optical networking world with its announcement of an all-optical switch based on parent company Hewlett-Packard's time-tested Bubble Jet technology. At the heart of their optical switch matrix is a thin piece of glass, upon which have been etched channels that in turn connect to fibers. Light signals traveling down the individual channels can be diverted to individual output fibers by the injection of tiny bubbles that deflect the signal in the desired direction. The bubbles are produced by heating liquid in the channels (Agilent won't say what the liquid is). This solution, which may sound like it was pulled from the pages of science fiction (isn't all of this optical stuff?), has been sold to Alcatel and will be used by the company in the design of optical switch platforms.

ALCATEL

Although Alcatel is well known for its SONET, SDH, and DSLAM hardware (among others), it is less well known as a manufacturer of intelligent optical networking equipment. However, their products are extremely capable and cover a wide range of solution areas in the optical networking domain. According to RHK, Alcatel holds dominant positions in total transport (20 percent), terrestrial and submarine DWDM systems, submarine systems (41 percent), and optical fibers. Clearly they are in an advantageous position in the marketplace. The company manufactures optoelectronic components, fiber, and optical transport systems.

ALIDIAN NETWORKS

Alidian delivers *Optical Service Network* (OSN) solutions in the metropolitan arena. The company's OSN4000 product line combines access, transport, switching, and DWDM into a single integrated platform for network simplicity. The products are fundamentally high-end, add-drop multiplexers that have the ability to scale from one to 32 OC-48c. One major selling point that the company touts is WavePack, which enables service providers to pack multiple services and protocols into a single wavelength, thus maximizing the efficient use of available optical bandwidth. The company is targeting a wide variety of applications, including private line replacement, cable, SONET/ SDH replacement, metro ring deployment, multi-tenant building buildouts, DSLAM transport, ISP and ASP connectivity, transparent LAN services, and connection services for storage area networks (SANs).

Amber Networks

Amber claims to be "Shaping the Optical Internet" with their ASR2000™ aggregation Service Router and their accompanying edgeSpan™ management system. The ASR2000 edge device enables service providers to deploy both switched and routed services from the same carrier-class device, thus further enabling the carrier's optical packet core by giving them the ability to aggregate TDM, frame relay, ATM, and IP. As a result, MPLS-enabled IP can start to become a reality.

Appian Communications

Appian delivers products that support the intelligent edge of the evolving optical network. By combining high-speed Ethernet with SONET, Appian has the ability to replace inflexible TDM-based local loops and provisioning bandwidth from 64 kbps to 10 Gbps. The company's Optical Services Activation Platform™ 4800 integrates Ethernet, SONET/SDH, and DWDM in a single platform, while AppianVista™, the company's services and element manager, ties it all together.

Astral Point Communications

Another metro market player, Astral Point has created an innovative approach to building and managing scalable, flexible TDM, broadband, Ethernet, and IP networks. The ON 2000 Optical Access Node, ON 5000 Optical Service Node, and Galaxy Distributed Control Plane, and On*set* Optical Network Management System provide element consolidation, a wide variety of interfaces, support for both ring and mesh topologies, bandwidth management, service provisioning, protection, and restoration.

Calient Networks

Calient manufactures MEMS-based photonic switches that rely on the company's patented SCREAM™ (*Scalable Control of a*

Rearrangeable Extensible Array of Mirrors) technology. Calient also offers embedded software intelligence and a comprehensive, Web-based network management system.

CENTERPOINT BROADBAND TECHNOLOGIES

The "centerpoint" of Centerpoint's success is their *Lightwave Efficient Network Solution* (LENS), a technique for delivering subcarrier multiplexing that permits up to 20 Gbps of traffic to be carried on a single wavelength. Targeted at regional and metropolitan applications, the LENS1220 is a protocol independent multiplexer that supports point-to-point, ring, and mesh architectures.

CIENA

Ciena made its name in 1996, by developing and shipping one of the first DWDM systems to enter the market. Today, the company offers a wide array of products including DWDM devices and optical switching platforms. Ciena's LightWorks™ architecture is an intelligent network platform developed by the company in concert with major service providers. It provides a suite of solutions that include the access, interoffice transport, and core regions of carrier networks.

Ciena's optical core switches include the MultiWave® CoreDirector™, and MultiWave® CoreDirector™ CI, an intelligent switching solution designed specifically for the smaller regional and metropolitan "regions" of carrier networks. The MultiWave® CoreStream™, and Sentry™ products are targeted at long-haul DWDM applications, while the MultiWave® Metro™, Metro One™, and Firefly™ products are for short-distance systems. Ciena also manufactures erbium-doped fiber amplifiers, optical add-drop multiplexers, and ON-Center™, a network management applications suite that provides deployment, monitoring, and service assurance services.

Ciena recently enjoyed the attention of the industry when the company announced that they would introduce 25 GHz and 12.5 GHz channel spacing on its MultiWave™ CoreStream™ DWDM products, enabling it to offer 16 Tbps systems in 2001.

CISCO SYSTEMS

Long considered the leader in the IP router world, Cisco recently entered the optical networking game through its acquisitions of SkyStone Systems, PipeLinks, Monterey Networks, Cerent, Internet Engineering Group, Pirelli Optical Systems, Growth Networks, and Qeyton Systems. They have entered the optical transport game with a suite of devices that satisfy the growing demands for both bandwidth and multiservice transport. The Cisco ONS 15303 and 15304 Optical Transport Extenders are premises-based optical transport devices that support TDM, Ethernet, SONET/SDH, and integrated routing functions. The ONS 15454, on the other hand, performs these same functions but adds integrated cross-connect functionality and data switching to satisfy the demands being placed on large service providers.

CORVIS

Corvis claims to be the pioneer of the all-optical network, and certainly a degree of truth to the claim exists. The Corvis CorWave optical product family includes the *Optical Network Gateway* (ONG), optical amplifiers, an *Optical Add-Drop Mux* (OADM) that is capable of adding and dropping individual wavelengths, and the Corvis Optical Switch, the first all-optical wavelength switch. Like the other products, the Optical Switch is designed to work seamlessly with CorManager, Corvis' comprehensive network management system.

As an optical solutions provider, Corvis has worked closely with a number of other vendors including Juniper Networks. During recent interoperability testing, the two companies successfully demonstrated the transmission of 10 Gbps traffic between Corvis and Juniper hardware. In an equally impressive trial, Corvis and Williams Communications recently demonstrated the ability to transmit optical signals 20,000 miles across the Williams network without the need to regenerate the signals.

CRESCENT NETWORKS

Crescent Networks has developed *Dense Virtual Routed Networking* (DVRN), which dramatically accelerates the delivery of IP traffic. In the same way that DWDM has revolution-ized optical networking, Crescent Networks' DVRN will change the face of IP service. DVRN uses secure, dynamic virtual rout-ing combined with QoS to offer public network services that emulate private line and dynamic networking. Thus a service provider can offer the best qualities of a private network and the dynamic Internet.

EXTREME NETWORKS

Extreme Networks targets the metropolitan and access markets with an array of products. The company's stackable Summit switches, Alpine Ethernet switches, and BlackDiamond core switches are designed to meet the constantly changing business requirements of network service providers by offering scalable speed, bandwidth, network size, and quality of service.

Within each Summit, Alpine, and BlackDiamond device is a non-blocking switch that offers wire-speed Layer 3 and Layer 2 performance. To guard against demand fluctuation, the back-plane capacity of Summit, Alpine, and BlackDiamond is greater than the aggregate of all ports.

All Extreme Networks switches are designed to work with and are shipped with the ExtremeWare standards-based software suite. ExtremeWare relies on industry-standard protocols to ensure interoperability with legacy switches and routers, as well as QoS for bandwidth management and traffic prioritization.

GEYSER NETWORKS

Responding to the market demand for converged solutions, Geyser offers the OSM4800™, a metropolitan network platform based upon the company's FlexBand™ technology that com-bines DWDM, DCS, SONET ADM, a QoS-aware IP router, an MPLS router, an Ethernet switch, an ATM and frame relay traf-fic aggregation device, and an *integrated access device* (IAD) in a single, small footprint chassis. Their MetroView™ network manager system allows for comprehensive provisioning and service management functions, and enables the service provider to offer highly granular control of network bandwidth, which can be allocated dynamically as TDM-like services.

LUCENT TECHNOLOGIES

Lucent Technologies covers the waterfront with an impressive array of optical networking hardware for the long haul, metropolitan, and local network areas including TDM multiplexers, DWDM equipment, cross connects, and terabit routers. Their TDM products, sold under the WaveStar™ name, include not only traditional SONET and SDH technology, but 2.5 Gbps, 10 Gbps, and 40 Gbps multiplexers, with higher bandwidth solutions on the horizon. These products interface with DWDM products from the *Optical Line System* (OLS) DWDM product line, including the OLS 400G and OLS800G. Higher bandwidth DWDM solutions are already in the works and will be released soon.

Equally impressive is the company's switch and router lineup. Their acquisition of Nexabit added the N64000 terabit router to the company's product line. On the cross-connect side of the house, the Bandwidth Manager provides an integrated broadband, wideband, ATM, and IP core network solution that is modular and highly scalable.

The LambdaRouter ranks among the most impressive devices in the Lucent lineup. An all-optical cross connect switch, the LambdaRouter is capable of transporting as many as 256 simultaneous wavelengths, soon to be scaled to 1,024. It is a carrier-class, full-duplex device that is totally protocol-agnostic. Based on MEMS technology, the LambdaRouter has recently been chosen by fiber network provider Global Crossing to serve as the interface point for the provisioning of wavelength-based services and as the wavelength cross-connect mechanism in their submarine network.

Metropolitan DWDM products are now available through Lucent's acquisition of Chromatis, which offers multiservice metropolitan access. DWDM has not been particularly successful in the metropolitan arena, but thanks to Chromatis' Selective WDM™ technology, service providers can now offer DWDM only where they need it on specific segments of a metro optical ring. It also permits sites on the ring that are not currently using DWDM to easily add it in the future as the need arises. The products support high-bandwidth ATM trunking services, wholesale wavelength provisioning, and cable system transport. Element and network management systems complete the product line.

MARCONI

Marconi has a long and distinguished background in network engineering, which it has now applied to the optical networking domain. Its DWDM and SONET/SDH equipment are part of an integrated network system called SmartPhotoniX, an integrated end-to-end optical networking solution that includes a wide variety of options including element and network management.

MAYAN NETWORKS

MAYAN plays at the edge of the metro network with the MAYAN Unifier, a next-generation metro access platform. The device aggregates, routes and switches TDM, frame relay, IP, and ATM traffic across the network, and is targeted at CLECs, ILECs, IXCs, and ISPs. It is fully compatible with legacy technology installations and offers a converged solution for service providers.

NORTEL NETWORKS

Nortel's strategy is to build upon its powerful position in the 10 Gbps marketplace, a market in which it acquired leadership by being the first entrant to establish an OC-192 beachhead. The company has many of the same urgencies as Lucent technologies, including a desire to move into the IP-based terabit router market and provide system-wide network management. Nortel's optical networking product line is called OPTera, which provides a remarkably complete collection of products including a diverse TDM family of 2.5, 10, and 40 Gbps multiplexers, a terabit IP router and packet core switch in the form of the Versalar 25000, a well-developed DWDM line, an optical cross-connect switch thanks to the company's acquisition of MEMS-manufacturer Xros, and a number of other, equally impressive products that offer a complete optical solution to customers. The OPTera Metro multiservice platform series is a ring-based, 80 Gbps DWDM system targeted at the metropolitan transport market. The OPTera Long Haul 1600 aggregates multiple channels across a 10 Gbps backbone. Their associated amplification system can take single fiber capacity as high as 1.6 Tbps. The OPTera Connect Connection Manager, their optical cross-connect system, is a high capacity MEMS-based system, and all are managed by Preside serviceware, the company's service management package.

OCULAR NETWORKS

Ocular's focus is on the metropolitan region, as evidenced by their *Metro Business Access Architecture*™ (MBA), which combines TDM, ATM, and IP traffic across the same switched fabric. Based on specially designed *Application-Specific Integrated Circuits* (ASICs), the MBA platform can dynamically allocate both switching and transport capacity. They are the first vendor to integrate diverse native traffic types across a single switched metro fabric. The company's *Unified Network Interface Technology* (UNIT)™, a combination of hardware and software, enables traffic to be classified for QoS control, while Ocular's Dynamic Bandwidth Sharing Protocol™ allows for the dynamic allocation of transport bandwidth based on traffic characteristics.

OPTXCON

OptXCon has entered the optical switching market with an LCD-based all-optical cross connect system. Their products are targeted at the enterprise and metro markets, and although relatively new are being well-received. OptXCon claims that its LCD technology is anything but routine. Backed by Adva AG Optical Networking, they believe that their solution will prove to be the long-term winner in the optical switching world because no moving parts wear out over time. They are not alone in this belief. Other startups, including Chorum Technologies and SpectraSwitch, Inc., have also come into the game with LCD-based solutions.

QUANTUM BRIDGE COMMUNICATIONS

Quantum Bridge specializes in optical access and edge networking, and was one of the first players to enter the PON game.

REDBACK NETWORKS
Redback Networks has risen to a position of some prominence by creating a family of highly integrated, multiprotocol platforms that enable carriers to provision, aggregate, and manage broadband, leased line, and dial customers from a single platform. Scalable to as many as a million subscribers per seven-foot rack, Redback's products come in several sizes and enable carriers to offer a broad range of value-added services. Redback also offers a highly integrated element, service, and network management system.

SYCAMORE NETWORKS™
Sycamore calls itself the founder of the intelligent optical network. Founded in 1998, the company rapidly rose to assume a position of prominence in the public optical networking domain. They offer access, transport, and core switching products that are designed around an intelligent software management model. Sycamore's optical product line includes the SN16000 optical core switch, the SN 3000 optical access switch, the SN 8000 intelligent optical transport platform, and SILVX™, optical network management system. Network management is considered by many to be one of Sycamore's greatest strengths in that it is one of the few fully integrated optical network management systems.

TRELLIS PHOTONICS
First there was MEMS. Then, Agilent entered the all-optical switch game with its bubble jet technology. Next came Gooch and Housego with its resonant crystalline lattice technology. And now comes Trellis Photonics with its Intelligent Lambda Switch, based on electroholography, another technique that seems to be just this side of magic—read on.

Trellis' switch relies on an array of 2×2×1.5 mm crystals of potassium lithium tantalate niobate, inside of which are written holograms by incident laser light. The hologram behaves like a Bragg filter, reflecting only certain specific wavelengths of light. One key difference is that the hologram is electrically energized. When voltage is applied to the KLTN crystals, the crystal becomes reflective; otherwise it is—well, crystal clear. An 8×8 switch (two input fibers, four output fibers) contains 64 crystals, while a larger switch (240×240) contains 1440 crystals. Trellis claims that the switches are massively scalable, and in fact expect to announce 3840-port by 3840-port arrays late in 2001. The good news about the technology according to Trellis, is that no moving parts exist and the reflective nature of the crystalline lattice is about 95 percent efficient. One potential downside is that it requires a relatively high voltage to energize the crystals (about 100 volts). Another is the inherent difficulty in cross-connecting wavelengths, but Trellis feels that they will overcome both shortcomings soon.

ULTRA FAST OPTICAL SYSTEMS

This company has an intriguing technological proposition. Many industry pundits believe that SONET will reach the limits of its transport lifetime at or around 40 Gbps. In response, Ultra Fast Optical Systems has announced *Optical Time-Division Multiplexing* (OTDM), a technology that is similar to SONET in that it has the ability to transport lower-speed services in high-speed channels. However, it differs from SONET in that it is an all-optical solution, which enables it to operate at extremely high speeds. According to a paper presented at the *European Council on Optical Networking* (ECOC 2000), OTDM can reach speeds as high as 640 Gbps—and that's one wavelength! OTDM interworks seamlessly with DWDM to deliver terabits of bandwidth.

ZAFFIRE, INC.

Zaffire targets the metropolitan and regional markets with what it calls the first packet-enabled, dense wavelength division multiplexing optical solution. Zaffire uses Fractional Wavelength™ technology in combination with multiprotocol label switching MPLS to control both bandwidth utilization and quality of service. The Z3000 OSP is designed to handle the specific requirements of metro and regional optical transport. The system supports a broad variety of traffic types and can scale to 256 wavelengths per fiber pair for a total transport speed of 2.5 Tbps.

FIBER MANUFACTURERS

Fiber manufacturers are a small family, with Corning and Lucent leading the parade. However, other well-known names are members of the family including Pirelli, Alcatel, Boston Optical Fiber, FiberCore, Furukawa, and Fujikura. This is a fiercely competitive market segment, currently plagued by a worldwide shortage of manufacturing capacity.

Two of the biggest challenges faced by the manufacturers of optical fiber are the tendency of optical signals to disperse as they are propagated down the fiber, and the tendency of the signal to weaken over distance (attenuation). As we discussed earlier, it is a well-known characteristic of optical fiber that minimal dispersion occurs at a wavelength of 1310 nm, while minimal attenuation occurs at 1550 nm. The newest forms of optical fiber, called dispersion shifted fiber, actually move the minimal dispersion out to 1550, causing both dispersion and attenuation minimums to occur at the same points. As was discussed in the last section, this is done by "doping" the fiber with elements such as germanium (doping is simply the process of embedding molecules of the dopant in the silica matrix of the fiber during the manufacturing process).

A number of players are active in the fiber game. Let's examine them and their products.

ALCATEL

Alcatel was the first company to commercially offer Raman amplification in its fiber products. Their TeraLight™ optical fiber permits close wavelength spacing, high bit rates and is ideally suited for long haul networks. The company claims a number of world records with regard to ultra long-haul transmission. In one trial, they transmitted 80 channels over a distance of 3,000 kilometers, with each channel operating at 10 Gbps. In another trial, they transported 32-40 Gbps channels over a 250 kilometer unrepeatered span; and in 1999, were able to achieve a 12,000 kilometer non-regenerated transmission between the U.S. and China.

Boston Optical Fiber

Boston Optical Fiber is the primary supplier of plastic optical fiber. Although plastic fiber is not considered to be a serious contender in the high-speed data transport world, it does have a place in the lineup. It is significantly cheaper than glass optical fiber, and in many circumstances more durable. According to the company, their POF product can transport data at 300 Mbps over distances as great as 100 meters, and they expect to release a graded index version soon that will transport signals at multi-gigabit speeds over the same distance.

Corning Optical Fiber

Corning produces a wide array of optical fiber products, both single and multimode. On the single mode side of the house, their *Large Effective Area* (LEAF™) product is ideally suited for long haul transport involving high-speed data traffic. MetroCor™ is an all-optical DWDM-focused metropolitan product, while SMF-28™ is considered by many to be the standard for regional, local, metropolitan, and CATV applications. Corning also manufactures an array of fiber specifically designed for the rigors of submarine installations.

In the multimode realm, Corning offers InfiniCor® CL™, designed for long reach, laser-based LAN installations, and InfiniCor®, specifically targeted at Gigabit Ethernet applications. Their 62.5/125 and 50/125 graded index multimode product, designed for local and wide area applications, rounds out the line.

Fibercore

Located in the UK, Fibercore is one of the oldest fiber manufacturing companies in the world. They produce a variety of products including a polarization maintaining fiber and a variety of rare earth-doped products, including erbium doped fibers.

Lucent Technologies
Lucent and Corning are the titans that dominate the optical fiber industry. The company's fiber products include the TrueWave® fiber line that offers a broad mix of products for many different environments. TrueWave® RS is an NZDF fiber designed to work in the third (C-Band) and fourth (L-Band) transmission windows. TrueWave® XL Submarine Fiber is a negative dispersion, large effective area fiber designed for long-haul submarine installations. TrueWave® *Submarine Reduced Slope* Fiber (SRS) is also a negative dispersion fiber designed for effective use with EDFA-based systems. AllWave™ is a low hydroxyl fiber that operates in the previously off-limits 1400 nm band and provides transport for low-cost CWDM systems, metro installations, and hybrid fiber-coax applications.

The Non-Fiber Fiber Companies
Many companies are attracting customers with the promise of low-cost freespace optical solutions. These companies are using open air laser transmission between buildings to create low-cost metropolitan networks. Given that some current industry estimates claim that 80 percent of all business buildings are not served by fiber, freespace optics would seem to be a good alternative solution.

AirFiber
AirFiber's principal product is called OptiMesh™. Based on wireless optical technology, OptiMesh™ creates a mesh configuration between optical transceiver nodes that operates at 622 Mbps. Fully redundant, OptiMesh™ delivers five nines of reliability and is easy to deploy. Because of its mesh topology, full route redundancy can be achieved.

Terabeam
Terabeam provides wireless optical connectivity at 5, 10, and 100 Mbps. The company offers Internet connectivity as well as metro transport, and guarantees QoS through the use of MPLS.

THE SERVICE PROVIDERS

The service providers fall into five principal categories: the *Incumbent Local Exchange Carriers* (ILECs)/*Incumbent PTTs*; the *Competitive Local Exchange Carriers* (CLECs)/*City Carriers*; the *Interexchange Carriers* (IXCs); the cable operators (MSOs); and the so-called "Bandwidth Barons." All have installed massive fiber infrastructures, and many have moved to offer value added services over their own networks by acquiring content or partnering with content providers.

THE INCUMBENT LOCAL EXCHANGE CARRIERS (ILECS)/INCUMBENT PTTS

ILECs tend to be relatively homogeneous in terms of the products and services they provide. Their strengths lie in the access and transport business, at which they excel. However, because of the "Commoditization" of this business, their ability to maintain marketshare is diminishing. For the most part, they have grown by acquiring more of the same; witness Bell Atlantic's acquisitions of NYNEX and GTE, or SBC's acquisitions of Pacific Bell, Nevada Bell, Ameritech, and SNET. They have expanded their footprint, but have not done much to diversify their product and service offerings. In fairness, good reasons for this strategy exist. The ILECs have realized that with long distance relief pending, they must create a wide area presence for themselves. Their larger business customers are not necessarily local companies. Although they may have a local presence, they tend to be national or even global. If the ILECs are to become full service service providers, they must be able to serve those customers on an end-to-end basis, thus eliminating the need for intermediaries. Without a wide area data network, they cannot accomplish this. The fact is, the market is the ILECs' to lose. Many analysts believe that customers will buy all services from the local service provider, if the local service provider has the ability to provision them. The holder of the access lines rules. Consequently, much of the company convergence activity of late has revolved around acquisition of access lines. Consider Qwest's acquisition of USWest, or Global Crossing's acquisition of Frontier. On a slightly different level, AT&T's acquisition of TCI is clearly a gambit for local loops, and more will follow.

So, are ILECs a dying breed? Will they be brought down by the smaller, more nimble CLECs that are nibbling away at their longstanding customer bases? They do face some serious challenges. Their networks were designed around the idea that they would control 100 percent of the market, and are therefore not the most cost-effective resource in an open and competitive market. Other models are far more cost-effective than the ILECs' circuit-switched infrastructures. As a result, the ILECs are reinventing themselves a piece at a time, and are of course expanding their market presence in a variety of ways.

THE COMPETITIVE LOCAL EXCHANGE CARRIERS (CLECs)/CITY CARRIERS

The CLECs are similar to the ILECs in that they sell a commodity. They differ from the ILECs, however, because they can. They tend to sell the more lucrative products and avoid the markets and services that don't enjoy high returns. For example, many CLECs focus on residential voice customers, while others go after business customers. According to a study conducted by *Data Communications Magazine* in September 1998, out of approximately 500 CLECs (most of them in New York and California metropolitan areas), only 15 were going after major business customers at the time. Although the numbers have since increased, the disparity still exists to a degree. All CLECs are not created equal. Their business strategies and business plans for carrying out those strategies and satisfying customers vary dramatically from company to company. They are, however, good performers within their identified market niches. The CLECs are obviously after the same access lines that the ILECs want to protect. The ongoing convergence of medium-size CLECs such as NextLink and Intermedia are nothing more than positioning moves.

CLECs face significant obstacles by virtue of the fact that they are CLECs. As alternatives to the ILECs, they rely on interconnection agreements with them because they must have a collocation presence within the ILECs' central offices to provide service. The 1996 Telecommunications Reform Act mandates that before the ILECs will be allowed into the long distance market, they must demonstrate that they have opened their local market to competitors, enabling equal access to unbundled facilities such as local loops and certain services. CLECs often complain that although the ILECs have agreed to the stipulations, they are not particularly quick to respond to CLEC requests for interconnnection service and therefore have the ability to exert some control on the pace at which CLECs can enter their markets. Of course, some check and balances are in place, such as Section 251 of the 1996 Communications Act. This component of the law requires that ILECs sell circuits, facilities, and services to their competitors that are "at least equal in quality to that provided by the local exchange carrier to itself or to any subsidiary, affiliate, or any other party to which the carrier provides interconnection."

The ILECs control 90 percent of the access lines in the United States, so CLECs face a significant challenge. Some of them, however, have proven to be innovative players. Teligent and Winstar, for example, offer wireless local loops that are targeted at small to midsize businesses, many of whom feel as if they do not get adequate attention from the ILECs. The service requires line of sight, but Winstar claims to be able to serve 75 percent of the buildings in a downtown area with the technology, which is LMDS-based. Furthermore, many CLECs claim to be able to offer better, more customized service than their ILEC competitors. Most customers agree that the technology products sold by the ILECs and the CLECs are identical. The difference, they claim, is the way they deal with their customers. CLECs believe themselves to be more customer-focused, claiming that the ILECs are still plagued by legacy monopoly mentality. Whatever the case, some CLECs have initiated differentiation programs to help them garner the favor of customers such as payback plans for downtime, online real-time usage reports, and negotiated service level agreements.

Ultimately, the success of the CLECs relies on three critical success factors. Local number portability must be viable, functional, and available; operations support standards must be in place and accepted; and discounts for unbundled network elements from ILECs must be on the order of 50 percent.

Table 4-1 shown below illustrates the degree to which CLECs have deployed fiber.

THE INTEREXCHANGE CARRIERS (IXCs)

The *interexchange carriers* (IXCs) face the greatest challenge of all the players, but are also the companies demonstrating the most innovative behavior in the face of adversity. With companies like Qwest and Level 3 building massively "overcapacitized" fiber networks, bandwidth is becoming so inexpensive and so universally available that it is evolving to a true commodity. The margins on it, therefore, are dropping rapidly. Furthermore, the number of companies that have entered the long distance

TABLE 4-1   The Deployed Fibers Used in CLECs

COMPANY	ROUTE MILES
Level 3 Communications	14,000
AT&T Local Services	13,500
Digital Teleport, Inc.	11,000
WorldCom	10,000
MediaOne	10,000
ITC Deltacom	9,200
McLeodUSA	9,000
Adelphia Business Solutions	7,500
Time Warner Telecom	7,326
GST Telecommunications	6,808
Electric Lightwave	6,300
NTS Communications	6,000
CapRock Communications	5,500
Cox Communications	5,000
ICG Communications	4,499

Source: New Paradigm Resources Group (America's Network)

market has grown, as has their diversity. Although the legacy players (AT&T, Sprint, MCI) continue to hold the bulk of the market, a collection of power companies, satellite providers, and "bandwidth barons" have entered the game and are seizing significant pieces of market share from the incumbents.

In response, the IXCs are fighting back by diversifying. All have entered the ISP game, offering Internet access across their backbones at competitive prices. They have also bought or built local twisted pair access infrastructures, cable companies, satellite companies, and a host of others. Consider AT&T as an example. Beginning with their long lines division, the company has grown into a multifaceted powerhouse that now owns IBM Global Networks, TCI, Teleport, Excite, and @Home, to name a few. They are a long distance, local, wireless, ISP, portal, and cable company, and have aggressive plans to be a true, full-service telecommunications provider as they flesh out their strategy for market positioning. And while their recent reorganization changes the face of the company, they are still a key player.

As Table 4-2 below illustrates, the incumbent service

TABLE 4-2   Service Providers and Their Fiber Infrastructures

COMPANY	ROUTE MILES	FIBER MILES
AT&T	53,000	N/A
WorldCom	77,000	N/A
Sprint	30,000	3.4 million
Bell Atlantic	86,000	5.62 million
BellSouth	70,064	N/A
GTE	17,000	N/A
SBC	100,000	5 million
USWest (now Qwest)	N/A	2.3 million

Source: America's Network

providers, both local and long distance, have installed massive fiber infrastructures.

The Cable Companies

Similar to the ILECs, cable companies provide a largely homogeneous service that is under attack by a variety of alternative providers. Although they are striving to provide diverse services, the challenge to do so is daunting. TCI is in a good position through its alliance with AT&T. Others will enjoy similar relationships in the future.

The Bandwidth Barons

These companies used their access to right-of-way to build massive, global, optical transport networks. They include such heavyweights as Qwest, Level 3, Global Crossing, and Tyco. Their intention has always been to make bandwidth as inexpensive as possible in order to become the "carrier's carrier," selling bandwidth to everyone in huge, inexpensive quantities. Most of them started as commodity providers, but soon moved up the optical food chain as they added new services. Consider, for example, Qwest CyberSolutions, the joint venture company between Qwest and KPMG that provides global ASP services. Qwest's network architecture speaks to their commitment to provision massive amounts of bandwidth. Two diversely routed conduits house two cables each with 96 optical strands per cable. Each strand is further subdivided through DWDM into 40 individual wavelengths, each operating at OC-192. The result is a network that offers an almost unimaginable 154 Tbps of bandwidth, many times the capacity of AT&T's network. Again, the key to success in this market is to provide more than simply bandwidth.

Challenges and Directions

So who are these companies and what are the principal challenges that they face? To understand them it is first important to understand the networking environment in which they conduct business. Generally speaking, the optical network is broken into four major segments: the wide area or long haul domain; the metropolitan access domain; the metropolitan transport domain; and the metropolitan enterprise domain.

The wide area transport domain is dominated by DWDM-based bandwidth multiplication in both terrestrial and submarine applications. High-volume transport is key here. The bandwidth barons and Interexchange Carriers have a strong vested interest in this arena and might commonly deploy 40 or even 80-channel DWDM systems to accommodate the transport needs of their national and multinational customers.

THE LAST MILE

One universal truth in technology exists—the closer one gets to the end user, the more diverse and complex the network becomes. In the world of optical networking, this fact is equally true. At the network's edge, the local loop must support multiple access technologies and multiple standards including frame relay, ATM, and IP. In response, companies have emerged that focus on the creation of "wide spectrum" network management systems that not only handle multiple protocols and services, but are flexible, scalable, single-seat systems that reside at the periphery of the network, close to the customer, rather than in the shadowy recesses of the central office. These systems will support the simultaneous provisioning of broadband voice and data applications, and will help to realize the true promise of convergence.

So, will fiber ever reach the home or small office? Absolutely. However, certain caveats must first be satisfied such as the cost of optoelectronics, the dearth of optical interfaces on consumer equipment, network limitations, and regulatory issues. In the meantime, alternative solutions have emerged with varying degrees of success, such as *Hybrid Fiber/Coax* (HFC) architectures that take advantage of in-place, fully-functional wiring schemes. HFC is considered to be a relatively low-cost solution for extending the reach of fiber. In the last few years, however, a number of fiber-to-the-home projects have been initiated worldwide, providing optical local loop connectivity to more than 300 million potential lines. SBC Communications recently announced plans to spend more than

$6 billion on its broadband infrastructure, including Project Pronto, which includes 12,000 miles of fiber in metropolitan regions to support the deployment of high-speed DSL service.

Metropolitan access is characterized by the deployment of ring architectures, used for the aggregation and transport of lower speed traffic. For example, a carrier might deploy a 10 Gbps metropolitan ring throughout a large city, which would then interconnect to lower speed, 2.5 Gbps access facilities— either point-to-point circuits or rings. SONET/SDH, as well as DWDM, are key technologies in this environment. ILECs/Incumbent PTTs and CLECs/City Carriers are involved in this segment of the marketplace, where they wish to serve as peering points in the network, providing high-speed interfaces to multiple protocols, technologies, and companies.

The ILECs/Incumbent PTTs, on the other hand, face unfettered competition from all sides, but they remain in an enviable position. They control access to the bulk of the customers through their control of the access lines. Their primary goal is to add high-speed access as quickly as they possibly can while continuing to ensure that they can deliver on promises of QoS.

Some service providers have reinvented themselves as broadband access carriers, offering a wide array of high-bandwidth access options including DSL, cable modems, and wireless solutions. These broadband access providers face a different set of issues than traditional carriers. First, they do not have a great deal of experience managing high bandwidth access services, and have never seen the tremendous growth that currently characterizes the market. Second, they tend to be quicker and more nimble than their traditional counterparts, making technology decisions that are in the best interest of their customers based on economies of scale and the potential to generate added revenue in innovative ways. These companies must deploy the most current technologies, and must ensure scalability if they are to meet the growing demands of their customer base. Many believe that these companies may eventually "own" the customer as broadband access catches on. They are deploying high-speed architectures designed to support the requirements of telecommuters, remote office installations, wireless business access, Gigabit Ethernet interoffice communications, and regional ISPs. For example, a carrier might deploy an OC-48/STM-16 metro ring that provides transport for traffic that originates on DSL-equipped local loops, interconnecting remote workers with a corporate network elsewhere in the metro area. Another carrier might deploy an all-optical metro infrastructure to support the huge traffic increases between base stations that result from the deployment of third generation wireless services. As data access over cable becomes more common, a carrier might build a high-speed access network designed to aggregate and transport traffic between cable customers and an ATM backbone.

As wireless local loop technologies such as *the Local Multipoint Distribution Service* (LMDS) find their way into the business access domain, broadband metro carriers will roll out high-speed rings to satisfy the demands of these and other similar services. Similarly, a regional ISP, with a need to connect to multiple *Network Access Points* (NAPs) and database locations, may be served by a broadband metro carrier's 10 Gbps metropolitan ring that provides transport for traffic that originates on lower-speed services such as OC-48/STM-16, OC-3/STM-1, and traditional TDM services.

Finally, a metropolitan ring can be used to interconnect corporate LANs. Several vendors have deployed multiprotocol multiplexers that enable Gigabit Ethernet to be transported across a 10 Gbps ring—clearly an application with promise, given the widespread deployment of Ethernet technology.

Metropolitan transport is precisely what the name implies— that segment of the transport market that delivers the high-speed rings used to aggregate and move lower speed traffic between locations, or onward to a wide area transport environment. Finally, metropolitan enterprise is the realm of innovative access techniques designed to provide high-bandwidth solutions for businesses.

Key Players

Now that we have discussed the types of companies that make up the optical service provisioning business, we should make mention of a few notables that are helping to further redefine the industry. It should be clear to the reader that the incumbent ILECs, CLECs, and IXCs have done a great deal to bring the industry as far as it has come. However, a few players are approaching the challenge from different angles and should be examined.

360NETWORKS

360Networks is aggressively building a global optical network that will comprise 88,000 route-miles and will connect more than 100 cities on four continents.

BROADWING

Broadwing offers an all-optical network that operates at 2.4 Gbps, as well as a wide range of collocation, Web hosting, ASP services, IT consulting, and marketing, and communications support services through their Affinity Division.

FIBERWORKS

FIBERWORKS is a "carrier's carrier," operating in the metropolitan space. Their goal as a facilities-based carrier is to resolve the last mile bottleneck challenge by installing DWDM-equipped optical rings in critical metro areas and installing ducted fiber—anywhere from 144 to 864 fibers per duct depending on projected demand. Atlanta is the first city to be equipped by FIBERWORKS, but 14 other major cities will be added over the next year. FIBERWORKS offers bandwidth ranging from DS-1 to SONET speeds, as well as leased lambdas, collocation, Web hosting, infrastructure design, and deployment services.

GLOBAL CROSSING

Global Crossing is aggressively building the world's largest IP-based, all-optical network. The company's 101,000 route-mile network already serves five continents, 27 countries and more than 200 major cities. The company offers frame relay, ATM, and IP-based services, and is based upon DWDM-enhanced, self-healing ring architectures.

LEVEL 3 COMMUNICATIONS

Level 3 is building a globally-deployed, IP-based network that will focus primarily on the transport needs of the Web-centric business market. The network will be end-to-end IP and will offer local, metro, and long distance transport.

METROMEDIA

Metromedia offers managed optical network services including IP connectivity and hosted application services through its subsidiary, AboveNet Communications. In a recent acquisition, Metromedia purchased SiteSmith, a Web services provider that will help Metromedia round out its service offerings.

QWEST

Qwest's optical network offers 25,500 miles of 2.4 Gbps transport throughout North America, a network that is now being expanded to cover the rest of the world. Through its relationship with KPN, Qwest already provides transport service in 46 European countries.

SPHERA OPTICAL NETWORKS

Sphera is a metropolitan transport company. By placing gateways in major metropolitan areas, they offer DWDM-enhanced services including their Optical VPN, hub, and point-to-point networks.

YIPES

Yipes offers an innovative transport service for business traffic that is based upon Gigabit Ethernet. They offer Yipes WAN, Yipes MAN, and Yipes NET Internet Service, all of which provide high-speed transport within their individual service domains with speeds ranging from 1 Mbps to 1 Gbps. Because the interface to their network is native Ethernet, no special protocol conversion needs to be done. Bandwidth is scalable from 1 Mbps to 1 Gbps, and strong SLAs guarantee the service provided by the company. This company is raising eyebrows.

REINVENTING THE NETWORK

Of course, we must keep in mind that phrase now made famous by the telephone company, which says, "If it ain't broke, don't fix it." Contrary to the belief that many would like to have, SONET and SDH are far from dead. They are among the most widely deployed transmission systems, and are deeply embedded in long distance, enterprise access, and even metro transport and access networks. The current direction seems to indicate that these technologies will enjoy a long, slow decline to retirement, as more capable optical technologies replace them. However, this will take the form of a slow evolution, not a slash-cut revolution. As the demand grows for more flexible bandwidth allocation, the technology will see to it that the available bandwidth exceeds the demand. The network will grow and evolve to accommodate the changes in the user and application profiles, creating enormous opportunities for emerging optical transmission technologies. The downside, of course, is that this evolution brings with it a rise in complexity as additional network elements are added, placing greater demands on network management systems to ensure that they are up to the task of meeting the unwavering demands of an increasingly exacting and technically competent customer.

TOWARD A NEW NETWORK PARADIGM

The answer to this conundrum is to reduce the overall complexity of the network by lowering the number of network elements, replacing copper networks with optical, and improving the reliability of the network as a whole. These can be done, but in some cases they run contrary to the direction that the network has taken. For example, until recently, optical components were corralled in the physical layer as SONET or SDH devices, providing physical transport capability for ATM and frame relay networks that in turn provided switching fabric for layer three protocols such as IP. With the arrival of *Wavelength Division Multiplexing* (WDM) in the late 90s, an additional optical sublayer was created "below" SONET/SDH. At the end of the 90s, we saw the arrival of optical switching, which burrowed into the space between the physical layer (SONET/SDH) and the switching layer (ATM/frame relay). This resulted in a growth in complexity as the three-layer protocol stack became five.

Today, a move is afoot to collapse the stack again by evolving to an all-optical network. Current product offerings make it possible to deploy optical networks not only in the long-haul, but in the metro and access regions as well.

WHY THE EVOLUTION?

One of the principal driving forces behind the evolution to all-IP networks is *convergence*. Traditionally defined as a technology phenomenon, convergence has more recently been recognized as a troika comprising a technology component, a company component, and a services component. *Technology convergence* defines the vortex that is inexorably pulling the industry toward packet-based IP networks because it makes sense to do so. *Quality of Service* (QoS) protocols and techniques continue to evolve. As network devices evolve to be able to respond to QoS demands, services will move to the IP network because it provides a quality-centric, universal protocol substrate that will support all service types across a single network infrastructure. From the point of view of a service provider, this is an ideal construct because it means that they can eliminate the multiple networks, support services, and billing infrastructures that they currently operate in favor of a single converged transport environment.

Today the typical large service provider operates the *Public Switched Telephone Network* (PSTN), the frame relay network, the ATM network, the IP network, the ISDN and DSL overlay networks, and in some cases wireless and X.25 networks. For each of these they must have maintenance, installation, operations, accounting, configuration, provisioning systems, personnel, and network management centers capable of surveilling them. The fact that they communicate with each other and occasionally share resources and customers makes their management that much more complex.

Now, consider the advantage of the IP migration that is inexorably underway. If convergence is real and it becomes possible to migrate all of these disparate services to a single network infrastructure, the service provider will realize significant benefits as the business of providing network services undergoes massive simplification. Already, ATM providers offer a service called *Frame Relay Bearer Service* (FRBS) from their ATM networks. Customers requesting frame relay service are handed a frame relay interface, and their frames are transported into the network. There, they are converted to cells for wide area transport across the ATM backbone, and converted once again to frames upon delivery to the receiving interface. Does the customer know that their data was actually transported as ATM traffic? No. Does it matter? Absolutely not—provided the agreed-upon quality of service stipulated in the frame relay service level agreement is met. In this example, the service provider has the ability to essentially halve their network complexity by delivering two services from a single network.

If we take this to its ultimate conclusion, we find a network that is capable of delivering all services with equally appropriate QoS levels. The service provider truly has become a *service provider*, now capable of focusing on the demands of the customer rather than on the underlying technology.

However, more to the convergence phenomenon than IP implementation exists and other attendant technologies make possible the migration to a packet-based infrastructure. A second aspect of convergence is *company convergence,* defined by the merger and acquisition feeding frenzy that is currently underway in the telecommunications industry. Most companies have realized that although they are technologically very good at what they do, two other critical truths define them. First, more and more, customers don't care, and don't *want* to care, about the underlying technologies that make their services work. They are content to simply accept the services that the technologies make possible, and happy to know that the service provider will worry about the network, leaving them to worry about their own products and services. Second, as customer demands become more stringent, players in the telecommunications game have come to realize that they need more capability than they currently have. Given the pace of the industry they are in, they also know that they do not have the time to develop that capability in-house, so instead, they go out and buy it in the form of an acquisition, a merger, or a strategic alliance. Their overall set of capabilities, then, grows with minimal pain and at a pace that is acceptable to the customer.

The third component of the convergence troika is *services convergence*. Tied in tightly to the other two piece parts, service convergence speaks to the fact that the average customer is looking today for a single source for their telecommunications networking services. The fact that those services may derive from the combined capabilities of multiple smaller companies is immaterial. What matters is that the customer gets the products and services that they require to do business.

How does this tie together, then? Rather elegantly, actually. If a product or service provider properly assesses the needs of their customers, they can position themselves to be much more than simply a "box" provider.

Consider the following example. Acme Router Corporation sells high-quality hardware that supports the growing Internet industry. When its customers call to place orders, Acme responds quickly, shipping precisely what the customer asks for, always within the timeframe requested. Acme doesn't really know why they are called, but they are happy to be the vendor of choice. In this case, Acme is nothing more than a commodity provider. Chances are good that they are being chosen because they are the lowest-cost vendor, and because commodities are defined as products that are differentiable on price alone, this is probably not the best business model for Acme to engage in.

Now change the model slightly. Acme, now a proactive provider of router-based services, expends considerable effort to understand not only the business drivers of their own customers but also the forces that affect their *customers'* customers. By understanding what it is that drives the so-called third tier, Acme can anticipate the requests that its direct customers will have, and provide a solution when they call—or better yet, be able to call *them* with a proposal based on Acme's unique understanding of the customer's marketplace. If Acme understands what services the third tier will be looking for, then they understand what services they should be positioned to provide. Furthermore, if they know the services they should have in their pantheon of capabilities, then they know what technologies they must add to their collection to meet those requirements, and therefore know which technology companies they must ally with in order to round out their complete service offering. This is convergence at its best: a combination of technological capability, marketplace understanding, and an unrelenting focus on service with an understanding that it is service, not technology, that drives the telecommunications economy.

THE EVOLVING IP PROTOCOL MODEL

As the traditional voice-centric central office bends to the pressure of diverse quality of service requirements from heterogeneous traffic types, a new network architecture is evolving that satisfies these demands. Figure 4-5 comprises a four-layer protocol stack. The Internet Protocol, or IP, shown at the top, lies at the center of this great evolution. With the possible exception of the SS7 signaling protocols, IP is the world's most widely deployed protocol. It also provides us with the only truly universal addressing scheme in existence, and is embedded deeply in every network operating system deployed today that is of any consequence. It has become the focal point for such environments as call centers, where universal routing of multiple traffic types to a single operator is highly desirable. Prior to the introduction of IP and unified messaging, the ability to do that was complex, costly and people-intensive. Each operator required multiple phone lines, and the call center required call routing software that was complex and costly in its own right. IP, in concert with other Internet-derived protocols such as HTTP, enables tremendous simplification of the call routing algorithm. Consider the following example. To contact the author of this book using every possible business contact technique, you would need an office telephone number, a fax number, a home telephone number, an e-mail address, a cell phone

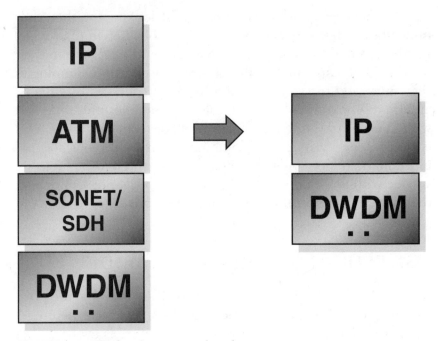

FIGURE 4-5   The four-layer protocol stack

number, a pager number, and so on. Chances are very good that the numbers would not be sequential, and would therefore be difficult to remember.

IP'S PROMISE

Now, consider the promise of IP's unified messaging concept. Instead of multiple unrelated numbers, the author could be contacted over an IP network in every possible way by typing

- *Steve@office.ShepardComm.com*
- *Steve@fax.ShepardComm.com*
- *Steve@home.ShepardComm.com*
- *Steve@mail.ShepardComm.com*

- *Steve@cell.ShepardComm.com*
- *Steve@pager.ShepardComm.com*

Clearly, this is a far simpler way to operate a business or call center, and dramatically simplifies the contact process for the customer. IP, then, provides the global addressing and universality required to make this possible.

Today, IP has a single drawback that is something of a showstopper. Because it was originally designed for the routing of connectionless, delay-insensitive data traffic across a packet network, it does not provide adequate quality of service (QoS) granularity to the broad range of services that it is now being asked to deliver. IP is something of a proletariat protocol in that it treats all traffic equally. By and large, in its *native mode*, it is incapable of discriminating between high and low-priority packets. This, of course, is a problem, because the diverse nature of traffic today requires a variety of QoS levels if the service provided by this single network fabric is to sell. Several options are either available or under development to accomplish this.

IP VERSION 6 (IPV6)

The first of these is the next generation of IP, known as IP Version 6 (IPv6). In IPv6, the protocol header has been redesigned to provide space for specific bytes that can be used to indicate the quality of service parameters required for each packet so that network routers can handle them accordingly. However, IPv6 is far from ready to be commercially deployed, and although it has been tested and is being trialed today, its widespread deployment is still a bit over the horizon.

TAG SWITCHING

A second method is to use a technique called *tag switching*. Originally developed by Cisco for quality control in large router networks, tag switching precedes each packet with an additional field, called a "tag," which contains quality of service requirements that network routers can take into account as they make routing decisions. Tag switching is a very capable technique, but has the drawback of being proprietary—it only works on Cisco routers. In response, an "open," vendor-independent form of tag switching was developed called *Multiprotocol Label Switching* (MPLS).

MULTIPROTOCOL LABEL SWITCHING (MPLS)

When establishing connections over an IP network, it is critical to manage traffic queues to ensure the proper treatment of packets that come from delay-sensitive services such as voice and video. In order to do this, packets must be differentiable, that is, identifiable so that they can be classified properly. Routers, in turn, must be able to respond properly to delay-sensitive traffic by implementing queue management processes. This requires that routers establish both normal and high-priority queues, and handle the traffic found in high priority routing queues faster than the arrival rate of the traffic.

MPLS delivers QoS by establishing virtual circuits known as *Label Switched Paths* (LSPs) that are built around traffic-specific QoS requirements. Thus, a router can establish LSPs with explicit QoS capabilities and route packets to those LSPs as required, guaranteeing the delay that a particular flow encounters on an end-to-end basis. It's interesting to note that some industry analysts have compared MPLS LSPs to the trunks established in the voice environment.

MPLS uses a two-part process for traffic differentiation and routing. First, it divides the packets into *Forwarding Equivalence Classes* (FECs) based on their quality of service requirements, and then maps the FECs to their next hop point. This process is performed at the point of ingress at the edge of the network. Each FEC is given a fixed-length "label" that accompanies each packet from hop to hop. At each router, the FEC label is examined and used to route the packet to the next hop point, where it is assigned a new label.

MPLS is a "shim" protocol that works closely with IP to help it deliver on QoS guarantees. Its implementation will enable the eventual dismissal of ATM as a required layer in the multimedia network protocol stack. And although it offers a promising solution, its widespread deployment is still a ways in the future because of the logistics of deployment.

ASYNCHRONOUS TRANSFER MODE (ATM)

The third technique, and the one that holds the greatest promise today, is found one layer down the protocol stack. ATM provides granular quality of service control through the capabilities of its *Adaptation Layer* (AAL).

The ATM Adaptation Layer at the ingress switch examines customer traffic as it arrives at the switch and then, based on the nature of the traffic, classifies it according to its QoS requirements. The requirements are based on three parameters: whether the traffic is connectionless or connection-oriented; whether it requires a fixed or variable bit rate; and whether or not an explicit timing relationship exists between the sending and receiving devices. Once these have been determined, the ATM ingress switch assigns a service class to the cells that make up the traffic stream and transmits them into the network, knowing that ATM's highly-reliable connection-oriented transport architecture and each switch's ability to interpret and respond to the assigned service class will ensure that the QoS mandate of the sending device will be accommodated on a network-wide basis. Thus, ATM provides QoS today.

Finally, we arrive at the lowest level of the protocol stack, where we encounter DWDM. DWDM offers massive bandwidth multiplication capability, and is in widespread use today. DWDM is a form of frequency division multiplexing, operating in the infrared domain, that enables multiple wavelengths of light to be simultaneously transmitted down the same fiber. This significantly increases the available bandwidth of the fiber and provides a cost-effective bandwidth multiplication solution to the provider.

Nothing is new about DWDM. In fact, it relies on technology that has been in widespread use throughout the network since the 1960s in the form of frequency division multiplexing, the technique of dividing a broad swath of spectrum into chunks and assigning each chunk to a different customer. As a technique for the facile multiplication of bandwidth, DWDM is a technological hero.

Of course, the success of WDM is more than simple multiplexing. Fundamental to its success was the ability to eliminate the optical-to-electrical-to-optical conversion that was necessary for switching, multiplexing, and signal regeneration—a process analogous to amplification in the analog transmission world. The first important accomplishment that led to this simplification was the development and widespread deployment of the all-optical amplifier.

Optical amplification, explained earlier, is the direct result of a sublime understanding of quantum physics. *Erbium-Doped Fiber Amplifiers* (EDFA) amplify signals in the optical domain, completely eliminating the O-E-O conversion that must normally take place.

EDFA is the technology that makes Wavelength Division Multiplexing commercially possible. By eliminating the need for electrical to optical conversion, the promise of the all-optical network can begin to be realized.

Before WDM became commercially available, optical transmission systems were for the most part limited to the transmission of a single wavelength per fiber, thus limiting the bandwidth of that fiber rather substantially by today's measure. As optical networking techniques continued to advance, however, this limitation became a non-issue. The original WDM systems developed by Lucent Technology's Bell Laboratories had the ability to transmit as many as four wavelengths of light down a single fiber. Today, DWDM systems routinely carry as many as 160 different wavelengths per fiber by assigning a different frequency, or color, to each stream of information. Individual lasers operating at different wavelengths transmit the information into the fiber, thus allowing enormous bandwidth to be offered from a single fiber.

At the time of transmission, the optical signal is amplified at the ingress point, after which it enters the fiber. Depending upon the nature of the fiber itself, the signal is then amplified every 40 to 60 miles to overcome the inevitable weakening of the signal that occurs over distance.

Using DWDM, service providers have the ability to add bandwidth to an existing fiber-optic network without having to engage the services of a backhoe. This becomes important when the costs involved are examined. As was mentioned earlier, the cost to add fiber to an existing network as a way to increase available bandwidth can be as much as $70K per mile. On the other hand, the addition of DWDM electronics at the endpoints to accomplish the same thing can cost as little as $12–20K—a significant difference. Much of the cost, of course, is labor, the requirement for which is dramatically reduced when the need for outside plant work is eliminated. This technique is often referred to in the industry as the deployment of "virtual bandwidth," because physical resources have not been added to bring about the improvement in the network that has taken place. In the same way that EDFA provided low-cost and highly-effective amplification to optical spans, other innovations have helped to reduce the complexity and expense of the migration from electrical to hybrid to all-optical networks. For example, optical switches, which use arrays of micro-mirrors, refractive bubbles in fluid-filled chambers, and the natural resonant frequency of certain types of crystals, make it possible to eliminate electrical switching elements. Tunable lasers eliminate the need for multiple lasers operating at specific frequencies, and reduce sparing requirements for service providers. Advanced and highly accurate optical filters provide channel separation in densely packed wavelength division systems, making it possible to dramatically expand the total bandwidth of an optical span. Management and monitoring systems, specifically designed for optical networks, make it possible to discretely manage these networks at highly granular levels, thus ensuring the ability to meet the requirements of customer service level agreements.

PROTOCOL ASSEMBLIES: PUTTING IT TOGETHER

IP, ATM, SONET/SDH, and DWDM represent a powerful and robust protocol stack, but in the minds of many industry pundits, are far from being the ultimate network design for the full-service transport fabric. One significant complaint that is often voiced about this four-tier stack is that it is highly overhead-intensive. True enough. IP, ATM, and SONET/SDH all add considerable overhead in the process of doing what they do. However, remember the adage: "If it ain't broke, don't fix it." Although some truth to this phrase exists, many argue that the model *is* broken, and can be significantly improved.

Consider what happens to customer traffic as it enters the IP-based network that we describe here. The stream of data enters the ingress router where it is chopped into pieces. To each piece is attached a header that contains information used to route the packets from the source to the destination. This header, although substantial, is usually inconsequential because IP packets tend to contain thousands of bytes of payload (user data).

The IP packets are then handed down to the ATM layer, where they are further segmented into 48-octet pieces. Each is given a five-byte header to form ATM cells. Now, the overhead in the header becomes significant—approximately 10 percent of the cell is overhead.

The cells are then handed down to the SONET/SDH layer, where they are packaged in frames for transport across the optical network. Each frame has embedded within it additional overhead, to the tune of about five percent of the frame.

It should be clear to the reader that some rather serious downsides exist to this four-layer stack that we envision. First of all, IP as it exists today, while a good protocol for universal networking, does precious little to guarantee the integrity of the user's data. ATM, for all its capabilities, is not really ideal for anything. It is not the best scheme for the transport of voice; the PSTN has it beat hands-down. And it's not the best for video; a dedicated high-speed circuit is far better. And it certainly isn't the best solution for data transport; it's far too expensive, and other solutions are equally capable. ATM is often described as being the telecommunications equivalent of a duck. Ducks walk, fly, swim, and make bird sounds—none of them particularly well.

Furthermore, the overhead tax that IP, ATM, and SONET/SDH exact is significant. Many argue that it doesn't matter, because of the belief that we are entering a time when bandwidth will be so abundant that we can afford to waste it. However, building networks based on that belief is irresponsible and dangerous. The communications corollary to Parkinson's Law promises that we will find a reason to need that bandwidth, so exercising caution to be efficient is advice worth listening to. Service providers are already burdened with the legacy of SONET and SDH, which were designed in a time when those deploying them were monopolies and not terribly concerned with protocol efficiencies. Many find the two to be monolithic, overhead-intensive, and inefficient—a stand that is hard to argue with.

One effort that is afoot (and that will undoubtedly be successful) intends to collapse the four-layer stack to two, eliminating the ATM and SONET/SDH layers entirely by moving their responsibilities into the IP and DWDM layers, respectively. In other words, the responsibility for quality of service control would be moved upward to IP, while survivability and robustness would become the responsibility of DWDM. To accomplish this, several things must happen. First of all, IP must become capable of managing and carrying out the guarantee of QoS, without benefit of ATM's adaptation layer. This could be done in several ways. First of all, the *Type of Service* (TOS) bits in the IP header could be used in concert with the *Differentiated Services* (DiffServ) protocol to create and respond to multiple QoS levels. This would require universal implementation of DiffServ/TOS throughout the greater network.

Second, MPLS could be deployed. This is the current favorite solution.

Finally, network stalwarts are out there who believe that IPv6 will arrive and be widely implemented, making possible the very granular QoS management that its overhead makes possible. One way or another, the function will be migrated upward.

Equally important is the migration of SONET/SDH's responsibilities to guarantee survivability downward into the DWDM sublayer. Today, DWDM provides massive amounts of bandwidth through the expediency of frequency division multiplexing. By dividing the optical bandwidth into channels, total throughput can be multiplied many times over. However, native-mode DWDM does nothing to guarantee the integrity of the information that it transports. It serves as nothing more than a multi-stream fire hose. Consequently, SONET/SDH's automatic protection switching, self-healing ring support and embedded network management protocols are necessary if the integrity of the network is to be guaranteed.

A new technology will soon change that requirement. Known generically as Digital Wrapper, it encloses each wavelength's traffic in a low-overhead frame of additional data that enables DWDM to detect and correct errors using sophisticated forward error correction techniques, perform optical layer performance monitoring, and provide ring protection on a wavelength-by-wavelength basis. Thus, many of the functions traditionally provided by SONET/SDH will be assumed by DWDM, eliminating the requirement for yet another protocol layer. Thus, the collapse of the four-layer protocol model into a two-layer construct is possible and highly likely.

The result of this protocol evolution is that functions traditionally performed within the core of the network—aggregation, prioritization, policy enforcement, QoS, and concentration of traffic—can now be performed at the edge of the network in the customer provided equipment (see Figure 4-6).

A final responsibility that is migrating from the core to the edge is switching (and routing). Historically, this process has been centralized because the devices required to do it were large and inordinately costly. Today, thanks largely to advances in microelectronics, that process, as well as the signaling responsibilities that govern it, is being substantially moved into high-speed edge routers.

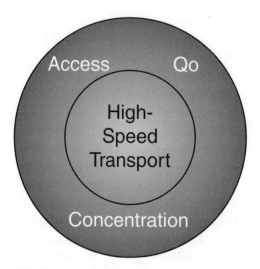

FIGURE 4-6   Hollowing out the network

So, what's left in the core? In reality, the only thing left is very high-speed transport. All setup processing, quality of service deliberation, traffic discrimination, and concentration are now performed by intelligent edge devices, while the core is left to provide extremely high-bandwidth transport. This is the domain of optical networking. This division of labor makes a great deal of sense and guides the design of modern high-speed networks as they struggle to meet the growing demands of bandwidth-hungry applications.

As the focus migrates from the core to the edge, one truth emerges: Enormous interest in the access region exists, and metropolitan transport is the hottest commodity out there. Referring back to the "network skin effect" described earlier, there's a lot of surface area to touch at the edge, and companies are rising up to do it. Optical networking is moving into all three regions of the metro space in a big way—metro core, metro access, and enterprise. Table 4-3 summarizes the key issues.

TABLE 4-3   Optical Networking Key Issues

	METRO CORE	METRO ACCESS	ENTERPRISE
Typical Distances	25–100 Km.	5–25 Km.	<10 Km.
Typical Applications	Point-to-point trunking, mesh connectivity	Optical add-drop, bandwidth management, ring networks, data center interconnect, SAN	
Systems in Place Today	OC12/48	OC3/12, ATM virtual path rings	Dedicated fiber
Upgrade Strategies	OC-192 or DWDM terminals for 40 Gbps capacity, optical switch/ cross-connect	Hybrid SONET/ ATM, ATM service access multiplexers, optical ADM	DWDM point-to-point or managed ring for virtual dark fiber
Benefits of DWDM	Fiber facility expansion, scalability, rapid restoration	Flexibility, rapid service provision, scalability	Efficient use of fiber capacity. Cost savings over dark fiber.

Source: Pioneer Consulting (Telephony)

ONE MORE TIME: PUTTING IT ALL TOGETHER

So what is it that customers want to do? At the highest possible level, customers, who include such vertical industries as residence, small business, finance, health care, education, government, and telecommunications, want to reduce their communications costs without sacrificing the quality or richness of the services that they depend on. These include voice, multimedia transport, interactive and distributive video, high-speed, QoS-aware data, diagnosis-quality image transfer, and a host of other traditional and non-traditional services.

The service providers that interface with these customers fall into a variety of categories, and each of them have specific issues and concerns that guide their efforts in the marketplace.

Incumbent local exchange carriers are taking extraordinary steps to upgrade their legacy, monopoly-flavored networks with more cost-effective architectures so that they can compete in their own newly competitive local markets while at the same time looking to move into long distance.

Competitive local exchange carriers, many of them facilities-based, must take advantage of high-bandwidth optical technologies as well as DSL if they are to compete cost-effectively with their incumbent competitors. Their market, which is primarily the mid-level business sector, can be well-accommodated by the CLECs as long as they can keep their provisioning costs under control.

The ILECs and CLECs have discovered a highly lucrative market segment in the form of the metropolitan transport market. This market, primarily serving multi-location corporate clients within a metro area, has become a major focal point and differentiation site for these companies as they struggle to battle each other for the market share. The metro market, characterized by the need for diverse, multiprotocol transport including Ethernet and other enterprise services, has ushered in demand for low-cost, service-centric systems capable of providing metro access and transport. The result has been the birth of system manufacturers that specialize in metro transport.

The third category of service provider is the interexchange carrier segment. This market, which includes such incumbents as AT&T, WorldCom, and Sprint, is beleaguered by incursions from the local exchange carriers on one end and the so-called "Bandwidth Barons" on the other. The IXCs face the same downward death spiral that the banking industry is currently going through. Competitors are offering lower-cost, high-quality alternatives, which cause an exodus of customers from the incumbent IXCs to the lower cost providers. This results in a revenue hemorrhage for the incumbent, which must either raise rates to compensate or reduce expenses, neither of which is a desirable solution. The fact is that pure bandwidth is fast becoming a commodity, and as we have already discussed, commodities are products that can be differentiated on price and price alone. To remain competitive, then, these companies must

take dramatic steps to make their networks and transport services as cost-effective as possible. Today, that is accomplished
through the deployment of optical technology. Optical networks
enable these companies to offer massive bandwidth at reasonable prices over a largely future-proof network.

The Bandwidth Barons, which include such notables as
Qwest, Level3, Global Crossing, Tycom, and 360Networks, are
playing for big stakes in the bandwidth game. Their networks,
made up of high-capacity linear facilities as well as massive,
redundant ring and mesh deployments, provide long distance
transport for local service providers. Of course, they too face
unbridled competition, and must take steps to differentiate
themselves. Qwest, for example, partnered with KPMG to
form Qwest CyberSolutions, an ASP-oriented enterprise that
offers a wide variety of data-oriented services including data
center offerings and hosted applications. Optical networking
figures prominently in their service architecture. If they are to
host mission-critical customer applications, they must be able
to guarantee not only bandwidth but access and survivability
as well.

Cable television providers are struggling to reinvent themselves as multiservice, one-stop providers, offering voice, data,
and interactive video services in addition to distributed content.
Unfortunately for them, they are struggling with the legacy of
an inferior network design and the market perception that their
networks are incapable of handling such a diversity of QoS-
dependent services. To counter this perception, cable providers
have spent billions retrofitting their networks with optical backbones and high-end digital compression technology to facilitate
the delivery of QoS-aware interactive services. Their efforts are
beginning to pay off.

The Internet service providers are experiencing similar problems, but for different reasons. The ISPs are the upstarts in this
business. A perception exists that because of their adolescence
they can't possibly provide high-quality service; and because
they deal with the terrible, undependable, insecure Internet,
they are not to be trusted. As a result, the ISPs are working to
add high-bandwidth capabilities to their own networks and

applications, and content to their services lineup to help overcome the image. Some are offering SLAs, which may contribute to the development of a more professional capable image.

Wireless providers also stand poised to benefit from the deployment of optical networking technology. As next-generation 3G/4G services arrive and make it possible to both receive and send broadband data from mobile devices, the upward pressure on transport networks at the edge and core will become enormous as all of that additional traffic impinges on the cloud. That traffic will require massive backbone buildouts if the network is to handle the volume.

Power companies, pipeline companies, and even railroads have also entered the bandwidth game, realizing as they have that their rights of way can serve as the fundamental underpinnings of a new set of value-added, revenue generating transport services.

What becomes obvious is that all of these companies are rapidly becoming dependent upon optical networking technologies. Furthermore, the key to the ability of service providers to truly provide diverse services lies in the availability of adequate bandwidth and the ability to quickly and easily provision that bandwidth on demand. That ability derives from the products created by the system manufacturers. So what do each of the service providers require in the way of optical system solutions?

The ILECs and CLECs, with their growing need to provide high bandwidth access to residence and metropolitan customers, require thousands of ports of SONET add-drop and terminal multiplexer capacity. Within the metro region, they may deploy DWDM in various segments of a ring, either in wholesale fashion throughout the ring or selectively, as is made feasible by Chromatis. They may also need to provide optical switching and cross-connect capability, as well as high-speed packet transport in the data environment.

The interexchange carriers, meanwhile, are into volume, and must therefore be able to move traffic quickly and efficiently. The optical technology upon which they are most focused is DWDM, although optical switching and amplification figure prominently in their architectures as well.

The Bandwidth Barons have similar requirements, although some of them, particularly those with submarine plant, have greater requirements for EDFAs and optical channel switching capability.

The cable, wireless, pipeline, ISPs, and railroads all have similar system requirements—devices used to build multiservice, protocol-independent access, and transport networks. This translates into ADMs and TMs, optical cross-connects, some specialized metro equipment, and DWDM.

Fiber, of course, is required by all of these companies. For short reach, low channel count WDM installations, 1310 nm fiber is perfectly acceptable. For long haul installations, especially those where industrial strength DWDM is to be deployed, NZDSF is used to counter the destructive effects of four-wave mixing.

This brings us to the last level in the services hierarchy, the component manufacturers. As the providers of subassemblies that make up the systems deployed by the service providers, we must consider what these devices are and where they fit. We will first examine a simple point-to-point network, followed by an add-drop ring. We will then examine switching and routing functions as appropriate.

At its simplest level, a point-to-point optical circuit comprises a laser source, a fiber transport, and a semiconductor receiver. In reality, the circuit is far more complex than that, as we will now see.

If the circuit is a single wavelength system, then a single laser provides the signal that is transported across the network. More likely, however, the system operates on a multichannel network as shown in Figure 4-7, in which case there will be multiple modulated sources (tunable lasers) feeding into a multiplexer, which will combine the various input signals for transport across the fiber. The lasers are most likely tuned with lithium niobate devices to provide highly stable signal sources. The fiber, which may be a long-distance facility, may have EDFAs and regenerators installed along the path. At the receive end is a demultipexer that feeds each of the signal components to a receiver, which converts the signal back to electrical for delivery to the end device. The receivers, which will either be APDs or PIN diodes, are highly sensitive receivers with a wide, dynamic range of reception.

Of course, it is also possible that the system is highly modern and has replaced the transmit lasers, lithium niobate modulators, multiplexers/demultiplexers, and receivers with transponders, which combine all of the functions into a single device.

If the network is a ring, very little changes, except that the ADMs that now appear on the network must provide add-drop capability in an intermediate fashion instead of purely at the end points of the circuit. Of course, to the ADM, the network is nothing more than a series of point-to-point circuits anyway, so logically little difference between the two architectures exists.

If we now introduce switching, we must add MEMS technology or some other fabric required to move wavelengths between input and output fibers. And of course, the nature of the transported traffic and the geography of the network itself will have significant bearing on the type of fiber selected.

We see, then, that the devices that appear in optical networks—whether they are access, metro, regional, long-haul, or submarine—are commonly used by all the systems. Although

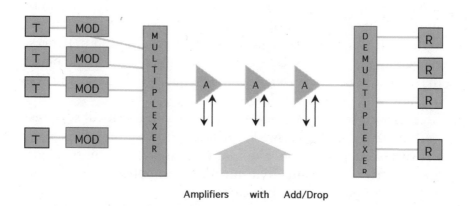

Amplifiers   with   Add/Drop

FIGURE 4-7   Network components

there may be some specialization that goes on, or specific devices that are only used in certain network regions, a certain commonality is at work here.

THE FUTURE

So what does the future hold for optical networking? Clearly no slowdown is on the horizon for the demand for bandwidth; and the diversity and richness of applications continues to grow. Mobility and multimedia will continue to drive massive volumes of traffic from the edge into the core; and QoS will continue to reign supreme for a very long time to come.

It is remarkable that technologies that continued to be innovative and future-proof a few years ago have now been relegated to the dusty halls of the legacy technology museum. SONET, once the great equalizer in the bandwidth game, is now considered by many to be old and decrepit.

Tomorrow's network will be IP-centric. However, if it is to work, something must be put into place to ensure that convergence of existing technologies and their capabilities with the new arrivals takes place to protect the embedded base that most carriers have.

Sometime around 1995, DWDM arrived and began to get a toe-hold in the modern network infrastructure. As it grew and service providers installed it in their networks, SONET began to show its limitations as it struggled to deal with the unpredictable traffic flows brought about by packet-based networks, especially the Internet. SONET was designed around the belief that network traffic patterns would remain predictable and stable, and that growth was largely linear.

It is well known that SONET provides four critical services: multiplexing, high-bandwidth service provisioning, discrete, network-wide performance monitoring, and survivability. These are non-trivial functions that must be accomplished, regardless of the technology that is in place.

Most network designers are working toward the all-optical Internet, which will completely bypass the SONET layer in favor of a flatter, photonic backbone. Today's modern network has a photonic physical layer, an ATM and SONET-based switching layer, and an IP-based routing layer. The all-optical network will eliminate the intermediate layer and connect the routing layer directly to the photonic layer. This will completely eliminate the need for most optical to electrical conversion, considered by many to be the most expensive part of the entire network infrastructure—as much as 40 percent of the overall network cost.

As terabit router manufacturers such as Lucent, Cisco, Nortel, and Juniper make good on their promises of systems with optical interfaces, they help to enable the all-optical backbone. The low cost of optical networking allows for the widespread deployment of mesh architectures, which, in combination with digital wrapper technology, effectively replace the restoration, provisioning, and management functions formerly performed by SONET. When this capability actually arrives, not only will the network be reinvented, but so too will the services and economics that drive the industry. As the cost of bandwidth plummets to the near-zero point, cost fears will disappear and innovation will kick in in ways that we can only imagine. Services will proliferate, solutions will evolve, and deployable value will result. The optical hierarchy of motivation will kick in as decisions begin to be made based upon the value of the deployed technology rather than the underlying features of the technology itself.

Maslow would be proud.

GLOSSARY

A

Absorption   A form of optical attenuation in which optical energy is converted into an alternative form, often heat. Often caused by impurities in the fiber, hydroxyl absorption is the best known form.

Acceptance Angle   The critical angle within which incident light is totally internally reflected inside the core of an optical fiber.

Add-Drop Multiplexer (ADM)   A device used in SONET and SDH systems that has the ability to add and remove signal components without having to demultiplex the entire transmitted transmission stream, a significant advantage over legacy multiplexing systems such as DS3.

Aerial Plant   Transmission equipment (including media, amplifiers, splice cases, and so on) that is suspended in the air between poles.

Amplifier   A device that increases the transmitted power of a signal. Amplifiers are typically spaced at carefully selected intervals along a transmission span.

Amplitude Modulation   A signal encoding technique in which the amplitude of the carrier is modified according to the behavior of the signal that it is transporting.

Analog   A signal that is continuously varying in time. Functionally, the opposite of digital.

Angular misalignment   The reason for loss that occurs at the fiber ingress point. If the light source is improperly aligned with the fiber's core, some of the incident light will be lost, leading to reduced signal strength.

Armor   The rigid protective coating on some fiber cables that protects them from crushing and from chewing by rodents.

Application-Specific Integrated Circuit (ASIC)   A specially designed IC created for a specific application.

Asynchronous   Data that is transmitted between two devices that do not share a common clock source.

Asynchronous Transfer Mode (ATM)   A standard for switching and multiplexing that relies on the transport of fixed-size data entities called cells which are 53 octets in length. ATM has enjoyed a great deal of attention lately because its internal workings enable it to provide *Quality of Service* (QoS), a much-demanded option in modern data networks.

Attenuation   The reduction in signal strength in optical fiber that results from absorption and scattering effects.

Avalanche Photodiode (APD)   An optical semiconductor receiver that has the ability to amplify weak, received optical signals by "multiplying" the number of received photons to intensify the strength of the received signal. APDs are used in transmission systems where receiver sensitivity is a critical issue.

Axis   The center line of an optical fiber.

B

Back Scattering   The problem that occurs when light is scattered backward into the transmitter of an optical system. This impairment is analogous to echo which occurs in copper-based systems.

Bandwidth   The range of frequencies within which a transmission system operates.

Baud   The *signaling rate* of a transmission system. This is one of the most misunderstood terms in all of telecommunications. Often used synonymously with bits-per-second, baud usually has a very different meaning. By using multibit encoding techniques, a single signal can simultaneously represent multiple bits. Thus, the bit rate can be many times the signaling rate.

Beam Splitter   An optical device used to direct a single signal in multiple directions through the use of a partially reflective mirror or some form of an optical filter.

Bending Loss   Loss that occurs when a fiber is bent far enough that its maximum allowable bend radius is exceeded. In this case, some of the light escapes from the waveguide resulting in signal degradation.

Bend Radius   The maximum degree to which a fiber can be bent before serious signal loss or fiber breakage occurs. Bend radius is one of the functional characteristics of most fiber products.

Bidirectional   A system that is capable of transmitting simultaneously in both directions.

Bragg Grating   A device that relies on the formation of interference patterns to filter specific wavelengths of light from a transmitted signal. In optical systems, Bragg Gratings are usually created by wrapping a grating of the correct size around a piece of fiber that has been made photosensitive. The fiber is then exposed to strong ultraviolet light which passes through the grating, forming areas of high and low refractive indices. Bragg Gratings (or filters, as they are often called) are used for selecting certain wavelengths of a transmitted signal, and are often used in optical switches, DWDM systems and tunable lasers.

Broadband   Historically, broadband meant "any signal that is faster than the ISDN Primary Rate (T1 or E1). Today, it means "big pipe" — in other words, a very high transmission speed.

Buffer   A coating that surrounds optical fiber in a cable and offers protection from water, abrasion, and so on.

Butt Splice   A technique in which two fibers are joined end-to-end by fusing them with heat or optical cement.

C

Cable   An assembly made up of multiple optical or electrical conductors, as well as other inclusions such as strength members, waterproofing materials, armor, etc.

Cable Assembly   A complete optical cable that includes the fiber itself and terminators on each end to make it capable of attaching to a transmission or receive device.

Cable Plant   The entire collection of transmission equipment in a system, including the signal emitters, the transport media, the switching and multiplexing equipment, and the receive devices.

Center Wavelength   The central operating wavelength of a laser used for data transmission.

Central Office   The central switching facility where most voice and data circuits terminate.

Chirp   A problem that occurs in laser diodes when the center wavelength shifts momentarily during the transmission of a single pulse. Chirp is due to instability of the laser itself.

Chromatic Dispersion   Because the wavelength of transmitted light determines its propagation speed in an optical fiber, different wavelengths of light will travel at different speeds during transmission. As a result, the multiwavelength pulse will tend to "spread out" during transmission, causing difficulties for the receive device. Material dispersion, waveguide dispersion, and profile dispersion all contribute to the problem.

Cladding   The fused silica "coating" that surrounds the core of an optical fiber. It typically has a different index of refraction than the core, causing light that escapes from the core into the cladding to be refracted back into the core.

Complimentary Metal Oxide Semiconductor (CMOS)   A form of integrated circuit technology that is typically used in low-speed and low-power applications.

Coating   The plastic substance that covers the cladding of an optical fiber. It is used to prevent damage to the fiber itself through abrasion.

Coherent   A form of emitted light in which all the rays of the transmitted light align themselves along the same transmission axis, resulting in a narrow, tightly focused beam. Lasers emit coherent light.

Concatenation   The technique used in SONET and SDH in which multiple payloads are "ganged" together to form a super-rate frame capable of transporting payloads greater in size than the basic transmission speed of the system. Thus, an OC-12c provides 622.08 Mbps of total bandwidth, as opposed to an OC-12, which also offers 622.08 Mbps, but in increments of OC-1 (51.84 Mbps).

Connector   A device, usually mechanical, used to connect a fiber to a transmit or receive device, or to bond two fibers.

Core   The central portion of an optical fiber that provides the primary transmission path for an optical signal. It usually has a higher index of refraction than the cladding.

Counter-Rotating Ring   A form of transmission system that comprises two rings operating in opposite directions. Typically, one ring serves as the active path while the other serves as the protect or backup path.

Critical Angle   The angle at which total internal reflection occurs.

Cross-Phase Modulation (XPM)   A problem that occurs in optical fiber that results from the nonlinear index of refraction of the silica in the fiber. Because the index of refraction varies according to the strength of the transmitted signal, some signals interact with each other in destructive ways. Cross-Phase Modulation is considered to be a fiber nonlinearity.

Cutoff Wavelength   The wavelength below which single mode fiber ceases to be single mode.

D

Dark Fiber   Optical fiber that is sometimes leased to a client that is not connected to a transmitter or receiver. In a dark fiber installation, it is the customer's responsibility to terminate the fiber.

Decibel (dB)   A logarithmic measure of the strength of a transmitted signal. Because it is a logarithmic measure, a 20 dB loss would indicate that the received signal is one one-hundredth its original strength.

Dense Wavelength Division Multiplexing (DWDM)   A form of frequency division multiplexing in which multiple wavelengths of light are transmitted across the same optical fiber. These DWDM systems typically operate in the so-called L-Band (1625 nm) and have channels that are spaced between 50 and 100 GHz apart. Newly-announced products may dramatically reduce this spacing.

Detector   An optical receive device that converts an optical signal into an electrical signal so that it can be handed off to a switch, router, multiplexer, or other electrical transmission device. These devices are usually either NPN or APDs.

Diameter Mismatch Loss   Loss that occurs when the diameter of a light emitter and the diameter of the ingress fiber's core are dramatically different.

Dichroic Filter   A filter that transmits light in a wavelength-specific fashion, reflecting non-selected wavelengths.

Dielectric   A substance that is non-conducting.

Diffraction Grating   A grid of closely spaced lines that are used to selectively direct specific wavelengths of light as required.

Digital   A signal characterized by discrete states. The opposite of analog.

Diode   A semiconductor device that only allows current to flow in a single direction.

Dispersion   The spreading of a light signal over time that results from modal or chromatic inefficiencies in the fiber.

Dispersion Compensating Fiber (DCF)   A segment of fiber that exhibits the opposite dispersion effect of the fiber to which it is coupled. DCF is used to counteract the dispersion of the other fiber.

Dispersion-Shifted Fiber (DSF)   A form of optical fiber that is designed to exhibit zero dispersion within the C-Band (1550 nm). DSF does not work well for DWDM because of Four Wave Mixing problems; Non-Zero Dispersion Shifted Fiber is used instead.

Dopant   Substances used to lower the refractive index of the silica used in optical fiber.

E

Edge-Emitting Diode   A diode that emits light from the edge of the device rather than the surface, resulting in a more coherent and directed beam of light.

Effective Area   The cross-section of a single-mode fiber that carries the optical signal.

Erbium-Doped Fiber Amplifier (EDFA)   A form of optical amplifier that uses the element erbium to bring about the amplification process. Erbium has the enviable quality that when struck by light operating at 980 nm, it emits photons in the 1550 nm range, thus providing agnostic amplification for signals operating in the same transmission window.

Evanescent Wave   Light that travels down the inner layer of the cladding instead of down the fiber core.

Eye Pattern   A measure of the degree to which bit errors are occurring in optical transmission systems. The width of the "eyes" (Eye Patterns look like figure eights lying on their sides) indicates the relative bit error rate.

Extrinsic Loss   Loss that occurs at splice points in an optical fiber.

F

Faraday Effect   Sometimes called the magneto-optical effect, the Faraday Effect describes the degree to which some materials can cause the polarization angle of incident light to change when placed within a magnetic field that is parallel to the propagation direction.

Ferrule   A rigid or semi-rigid tube that surrounds optical fibers and protects them.

Fiber Grating   A segment of photosensitive optical fiber that has been treated with ultraviolet light to create a refractive index within the fiber that varies periodically along its length. It operates analogously to a fiber grating and is used to select specific wavelengths of light for transmission.

Fiber-to-the-Curb (FTTC)   A transmission architecture for service delivery in which a fiber is installed in a neighborhood and terminated at a junction box. From there, coaxial cable or twisted pair can be cross-connected from the O-E converter to the customer premises. If coax is used, the system is called *Hybrid Fiber Coax* (HFC). Twisted pair-based systems are called *Switched Digital Video* (SDV).

Fiber-to-the-Home (FTTH)   Similar to FTTC, except that FTTH extends the optical fiber all the way to the customer premises.

Four Wave Mixing (FWM)   The nastiest of the so-called fiber nonlinearities. FWM is commonly seen in DWDM systems and occurs when the closely spaced channels mix and generate the equivalent of optical sidebands. The number of these sidebands can be expressed by the equation $1/2(n^3 - n^2)$, where n is the number of original channels in the system. Thus, a 16-channel DWDM system will potentially generate 1,920 interfering sidebands!

Fresnel Loss   The loss that occurs at the interface between the head of the fiber and the light source to which it is attached. At air-glass interfaces, the loss usually equates to about four percent.

Fused Fiber   A group of fibers that are fused together so that they will remain in alignment. They are often used in one-to-many distribution systems for the propagation of a single signal to multiple destinations. Fused fiber devices play a key role in *Passive Optical Networking* (PON).

Fusion Splice   A splice made by melting the ends of the fibers together.

G

Graded Index Fiber (GRIN)   A type of fiber in which the refractive index changes gradually between the central axis of the fiber and the outer layer, instead of abruptly at the core-cladding interface.

H

Hertz (Hz)   A measure of cycles per second in transmission systems.

Hybrid Fiber Coax   A transmission system architecture in which a fiber feeder penetrates a service area and is then cross-connected to coaxial cable feeders into the customers' premises.

I

Index of refraction   A measure of the ratio between the velocity of light in a vacuum and the velocity of the same light in an optical fiber. The refractive index is always greater than one and is denoted 'n.'

Infrared (IR)   The region of the spectrum within which most optical transmission systems operate, found between 700 nm and 0.1 mm.

Injection Laser   A semiconductor laser (synonym).

Intermodulation   A fiber nonlinearity that is similar to four-wave mixing, in which the power-dependent refractive index of the transmission medium enables signals to mix and create destructive sidebands.

Intrinsic Loss   Loss that occurs as the result of physical differences in the two fibers being spliced.

J

Jacket   The protective outer coating of an optical fiber cable. The jacket may be polyethylene, Kevlar®, or metallic.

Jumper   An optical cable assembly, usually fairly short, that is terminated on both ends with connectors.

L

Large Core Fiber   Fiber that characteristically has a core diameter of 200 microns or more.

Laser   An acronym for "Light Amplification by the Stimulated Emission of Radiation." Lasers are used in optical transmission systems because they produce coherent light that is almost purely monochromatic.

Laser Diode (LD)   A diode that produces coherent light when a forward biasing current is applied to it.

Light Emitting Diode (LED)   A diode that emits incoherent light when a forward bias current is applied to it. LEDs are typically used in shorter distance, lower speed systems.

Lightguide   A term that is used synonymously with optical fiber.

Linewidth   The spectrum of wavelengths that make up an optical signal.

Loose Tube Optical Cable   An optical cable assembly in which the fibers within the cable are loosely contained within tubes inside the sheath of the cable. The fibers are able to move within the tube, thus enabling them to adapt and move without damage as the cable is flexed and stretched.

Loss   The reduction in signal strength that occurs over distance, usually expressed in decibels.

M

Material Dispersion   A dispersion effect caused by the fact that different wavelengths of light travel at different speeds through a medium.

Microbend   Changes in the physical structure of an optical fiber caused by bending that can result in light leakage from the fiber.

Modal Dispersion   (See Multimode Dispersion.)

Mode   A single wave that propagates down a fiber. Multimode fiber enables multiple modes to travel, while single mode fiber enables only a single mode to be transmitted.

Modulation   The process of changing or *modulating* a carrier wave to cause it to carry information.

Multimode Dispersion   Sometimes referred to as modal dispersion, multimode dispersion is caused by the fact that different modes take different times to move from the ingress point to the egress point of a fiber, thus resulting in modal spreading.

Multimode Fiber   Fiber that has a core diameter of 62.5 microns or greater, wide enough to enable multiple modes of light to be simultaneously transmitted down the fiber.

Multiplexer   A device that has the capability to combine multiple inputs into a single output as a way to reduce the requirement for additional transmission facilities.

N

Near-End Crosstalk (NEXT)   The problem that occurs when an optical signal is reflected back toward the input port from one or more output ports. This problem is sometimes referred to as "isolation directivity."

Non-Dispersion Shifted Fiber (NDSF)   Fiber that is designed to operate at the low-dispersion, second operational window (1,310 nm).

Non-Zero Dispersion-Shifted Fiber (NZDSF)   A form of single mode fiber that is designed to operate just outside the 1,550 nm window so that fiber nonlinearities, particularly FWM, are minimized.

Numerical Aperture (NA)   A measure of the capability of a fiber to gather light. NA is also a measure of the maximum angle at which a light source can be from the center axis of a fiber in order to collect light.

O

OC-n   Optical Carrier level n, a measure of bandwidth used in SONET systems. OC-1 is 51.84 Mbps; OC-n is n times 51.84 Mbps.

Optical Amplifier   A device that amplifies an optical signal without first converting it to an electrical signal.

Optical Isolator   A device used to selectively block specific wavelengths of light.

Optical Time Domain Reflectometer (OTDR)   A device used to detect failures in an optical span by measuring the amount of light reflected back from the air-glass interface at the failure point.

P

Photodetector   A device used to detect an incoming optical signal and convert it to an electrical output.

Photodiode   A semiconductor that converts light to electricity.

Photon   The fundamental unit of light, sometimes referred to as a quantum of electromagnetic energy.

Photonic   The optical equivalent of the term "electronic."

Planar Waveguide   A waveguide fabricated from a flat material such as a sheet of glass, into which etched fine lines are used to conduct optical signals.

Plenum   The air handling space in buildings found inside walls, under floors, and above ceilings. The plenum spaces are often used as conduits for optical cables.

Plenum Cable   Cable that passes fire retardant tests so that it can legally be used in plenum installations.

Polarization   The process of modifying the direction of the magnetic field within a light wave.

Polarization Mode Dispersion (PMD)   The problem that occurs when light waves with different polarization planes in the same fiber travel at different velocities down the fiber.

Preform   The cylindrical mass of highly pure fused silica from which optical fiber is drawn during the manufacturing process. In the industry, the preform is sometimes referred to as a "gob."

Pulse Spreading   The widening or spreading out of an optical signal that occurs over distance in a fiber.

Pump Laser   The laser that provides the energy used to excite the dopant in an optical amplifier.

R

Rayleigh Scattering   A scattering effect that occurs in optical fiber as the result of fluctuations in silica density or chemical composition. Metal ions in the fiber often cause Rayleigh Scattering.

Refraction   The change in direction that occurs in a light wave as it passes from one medium into another. The most common example is the "bending" that is often seen to occur when a stick is inserted into water.

Refractive Index   A measure of the speed at which light travels through a medium, usually expressed as a ration compared to the speed of the same light in a vacuum.

Regenerative Repeater   A device that reconstructs and regenerates a transmitted signal that has been weakened over distance.

S

Scattering   The "backsplash" or reflection of an optical signal that occurs when it is reflected by small inclusions or particles in the fiber.

Synchronous Digital Hierarchy (SDH)   The European equivalent of SONET.

Self-Phase Modulation (SPM)   The refractive index of glass is directly related to the power of the transmitted signal. As the power fluctuates, so too does the index of refraction, causing waveform distortion.

Sheath   One of the layers of protective coating in an optical fiber cable.

Single Mode Fiber (SMF)   The most popular form of fiber today, characterized by the fact that it enables only a single mode of light to propagate down the fiber.

Soliton   A unique waveform that takes advantage of nonlinearities in the fiber medium, the result of which is a signal that suffers essentially no dispersion effects over long distances. Soliton transmission is an area of significant study at the moment, because of the promise it holds for long-haul transmission systems.

Synchronous Optical Network (SONET)   A multiplexing standard that begins at DS3 and provides standards-based multiplexing up to gigabit speeds. SONET is widely used in telephone company long-haul transmission systems, and was one of the first widely deployed optical transmission systems.

Source   The emitter of light in an optical transmission system.

Step Index Fiber   Fiber that exhibits a continuous refractive index in the core, which then "steps" at the core-cladding interface.

Stimulated Brillouin Scattering (SBS)   A fiber nonlinearity that occurs when a light signal traveling down a fiber interacts with acoustic vibrations in the glass matrix (sometimes called photon-phonon interaction), causing light to be scattered or reflected back toward the source.

Stimulated Raman Scattering (SRS)   A fiber nonlinearity that occurs when power from short wavelength, high power channels is bled into longer wavelength, lower power channels.

Strength Member   The strand within an optical cable that is used to provide tensile strength to the overall assembly. The member is usually composed of steel, fiberglass, or Aramid yarn.

Surface Emitting Diode   A semiconductor that emits light from its surface, resulting in a low power, broad spectrum emission.

T

Tight Buffer Cable   An optical cable in which the fibers are tightly bound by the surrounding material.

Total Internal Reflection   The phenomenon that occurs when light strikes a surface at such an angle that all of the light is reflected back into the transporting medium. In optical fiber, total internal reflection is achieved by keeping the light source and the fiber core oriented along the same axis so that the light that enters the core is reflected back into the core at the core-cladding interface.

Transceiver   A device that incorporates both a transmitter and a receiver in the same housing, thus reducing the need for rack space.

Transponder   A device that incorporates a transmitter, a receiver, and a multiplexer on a single chassis.

V

Vertical Cavity Surface Emitting Laser (VCSEL)   A small, highly efficient laser that emits light vertically from the surface of the wafer on which it is made.

W

Waveguide   A medium that is designed to conduct light within itself over a significant distance, such as optical fiber.

Waveguide Dispersion   A form of chromatic dispersion that occurs when some of the light traveling in the core escapes into the cladding, traveling there at a different speed than the light in the core.

Wavelength   The distance between the same points on two consecutive waves in a chain — for example, from the peak of wave one to the peak of wave two. Wavelength is related to frequency by the equation

$\lambda = c/f,$

where lambda (λ) is the wavelength, c is the speed of light, and f is the frequency of the transmitted signal.

Wavelength Division Multiplexing (WDM)   The process of transmitting multiple wavelengths of light down a fiber.

Window   A region within which optical signals are transmitted at specific wavelengths to take advantage of a propagation characteristic that occurs there, such as minimum loss or dispersion.

Zero Dispersion Wavelength   The wavelength at which material and waveguide dispersion cancel each other.

BIBLIOGRAPHY

ONLINE RESOURCES

Corning, Inc.	www.corningfiber.com
Agere Systems	www.agere.com
Agilent	www.agilent.com
Agility	www.agility.com
AirFiber	www.airfiber.com
Alcatel	www.alcatel.com
Appian Communications	www.appiancommunications.com
Astral Point	www.astralpoint.com
ATG Advanced TelCom Group	www2.callatg.com
Boston Optical	www.bostonoptical.com
Broadwing	www.broadwing.com
Chromatis	www.chromatis.com
Cisco	www.cisco.com
Colo.com	www.colo.com
Corvis	www.corvis.com
Crescent Networks	www.crescent.com
Extreme Networks	www.extremenetworks.com
Fiber Optics Online	www.fiberopticsonline.com
Fibercore	www.fibercore.com
FPN Magazine	www.fpnmag.com

Fujikura	www.fujikura.com
Furukawa Electric Co.	www.furukawa.co.jp/english
JDS Uniphase	www.jdsuniphase.com
Light Reading	www.lightreading.com
Lightwave Microsystems	www.lightwavemicrosystems.com
Lucent	www.lucent.com
Mayan Networks	www.mayannetworks.com
Nanovation Technologies	www.nanovation.com
NEC	www.nec-global.com
Network World	www.nwfusion.com/topics/optical.html
Nortel	www.nortelnetworks.com
Novalux	www.novalux.com
Ocular Networks	www.ocularnetworks.com
Optical Domain Service Interconnect (ODSI)	www.odsi-coalition.com
Optical Internetworking Forum	www.oiforum.com
Optical Society of America	www.osa.org
Optical Solutions	www.opticalsolutions.com
Optronics	www.optronics.com
Pluris	www.pluris.com
Polatis	www.polatis.com
Qwest	www.qwest.com
Redback Networks	www.redbacknetworks.com
Southampton Photonics	www.southamptonphotonics.com

Sphera Networks	www.spheranetworks.com
Startech	www.startech.com/fiberoptics/
Sycamore	www.sycamore.com
Tachion	www.tachion.com
Telecommunications Industry Association	www.tiaonline.org
Terabeam	www.terabeam.com
Ultra Fast Optical Systems	www.ufos.com
Yipes	www.yipes.com

BOOKS

Clarke, Arthur C. *How the World Was One: Beyond the Global Village*. Bantam; New York, 1992.

Evans, Philip and Thomas S. Wurster. *Blown to Bits: How the New Economics of Information Transforms Strategy*. Harvard Business School Press; Boston, Massachusetts, 2000.

Goff, David R. *Fiber Optic Reference Guide: A Practical Guide to the Technology—Second Edition*. Focal Press; Boston, 1999.

Goralski, Walter J. *ADSL and DSL Technologies*. McGraw-Hill; New York, 1998.

Goralski, Walter J. *SONET: A Guide to Synchronous Optical Networks*. McGraw-Hill; New York, 1997.

Goralski, Walter J. and Matthew C. Kolon. *IP Telephony*. McGraw-Hill; New York, 2000.

Hecht, Jeff. *Understanding Fiber Optics—Third Edition*. Prentice-Hall; Upper Saddle River, New Jersey, 1999.

Lanning, Michael J. *Delivering Profitable Value*. Perseus Books; New York, 1998.

Metz, Christopher Y. *IP Switching Protocols and Architectures*. McGraw-Hill; New York, 1999.

Minoli, Daniel and Emma Minoli. *Delivering Voice Over IP Networks*. John Wiley & Sons; New York, 1998.

Rackham, Neil. *Spin Selling*. McGraw-Hill; New York, 1998.

Shepard, Steven. *Telecommunications Convergence: How to Profit from the Convergence of Technologies, Services, and Companies.* McGraw-Hill; New York, 2000.

Stern, Thomas E. and Krishna Bala. *Multiwavelength Optical Networks: A Layered Approach.* Addison Wesley Longman; Reading, Massachusetts, 1999.

Tapscott, Don. *Growing Up Digital*. McGraw-Hill; New York, 1998.

ARTICLES

"SONET's Still Alive and Kicking." *America's Network*; November 15, 2000.

"Dense Wavelength Division Multiplexing: from ATG's Communications & Networking Technology Guide Series;" sponsored by *Ciena*. Available at www.techguide.com.

"Voice Over Frame Relay: A technical brief by the Frame Relay Forum," available at www.frforum.com/4000/voicetechbrief.html.

"Sphera Optical Networks Takes Advantage of Teledensity in LayerOne's Optical Transport Exchange." *Sphera Optical Networks* press release, November 22, 2000.

"Optical Switches go Acoustic." *Light Reading*; www.lightreading.com; June 20, 2000.

Notes from an interview with Walter Goralski, October 1999.

"A Fever Pitch in Optical Networking." *Business Week Online*; April 11, 2000.

"Optical Transport Market Booms." An article from RHK. No date shown. RHK Report on *Optical Transport Market* 1999.

"Cisco Systems Completes the Acquisition of Pirelli Optical Systems." Cisco press release, February 16, 2000.

Abreu, Elinor. "New Fiber in Old Trenches." *The Industry Standard*; April 24, 2000.

Allen, Doug. "Dial me a Lambda." *Network Magazine*; date unknown.

Aun, Fred and David Hakala. "Bet the House on Optical Networks. That's What Three Networking Giants are Doing. Should You?" *ZDNet Networking News*; July 13, 2000.

Biagi, Susan. "Fiber Fast and Furious." *Telephony*; June 5, 2000.

Branson, Ken. "Local Networks Grow Fiber as Bandwidth Demand Escalates." *X-CHANGE*; June 2000.

Cacal, Voltaire D. "The Evolution of Optical Networking." Notes from a presentation by RHK at the Optical Internetworking Summit, January 20, 2000.

Clavenna, Scott. "The Ultimate Backbone." *Light Reading*; October 9, 2000.

Davis, Christopher C. "Fiber Optic Technology and its Role in the Information Revolution." www.ece.umd.edu/~davis/optfib.html.

Diantina, Pablo. "Latin America Gets More from its Fiber." *FiberSystems International*; June/July 2000.

Dominguez, Alex. "Light May Break its Own Speed Limit." *Associated Press*; July 19, 2000.

Fairley, Peter. "The Microphotonics Revolution." *Technology Review*; July/August 2000.

Fitz, Jonathan G. "Take This Bandwidth and Shove it." *Telephony*; June 5, 2000.

Gilder, George and Richard Vigilante. "The Sounds of Silence." *Gilder Technology Report*, August 2000, Volume V, Number 8.

Gilder, George and Richard Vigilante. "The Microcosm Strikes at Cisco City." *Gilder Technology Report*, September 2000, Volume V, Number 9.

Gonsalves, Chris. "Metromedia Fiber to Buy Web Services Provider SiteSmith." *EWeek*; October 11, 2000.

HartMayer, Ron. "Components are the Secret of Interactive Networks." *FiberSystems International*; June/July 2000.

Heywood, Peter. "Tunable Filters Go Solid State." *Light Reading*; November 3, 2000.

Isenberg, David. "The Rise of the Stupid Network."

Jander, Mary and Marguerite Reardon. "Corning Hedges its Bets." *Network World Online*; September 14, 2000.

Jander, Mary and Marguerite Reardon. "Components Explosion." *Network World Online*; December 6, 2000.

Kenward, Michael. "Mirror Magic Ushers in the All-Optical Network." *FiberSystems International*; June/July 2000.

Khan, Nisa. "Manufacturers Focus on the 40 Gbit/s Challenge." *FiberSystems International*; June/July 2000.

Kim, Philip. "Managing Mega-Networks: What's Happening at the Edge?" *Communications News*; April 2000.

Lindstrom, Annie. "Taming the Terrors of the Deep." A supplement to *America's Network*.

Lindstrom, Annie. "Unmasking the Fiber Barons, Parts I and II." *America's Network*; March 1, 2000.

Lynch, Grahame. "The Next Killer Access Technology: Laser. Is it Reliable? Is it Safe?" *America's Network*; June 1, 2000.

McGarvey, Joe. "Clear Focus in Optical is on Startups." *Inter@ctive Week*; June 26, 2000.

Miller, Elizabeth Starr. "Lucent Catches Metro Fever." *Telephony*; June 5, 2000.

Nathan, Dr. Sri. "The Internet's Optical Future." *Communications News*; December 1999.

Raynovich, R. Scott. "Redback Completes IP ASICs." *Light Reading*; November 16, 2000.

Reardon, Marguerite. "More Terabit on the Way." *Light Reading*; November 22, 2000.

Rigby, Pauline. "Trellis Gets $25M for Holographic Switch." *Light Reading*; October 3, 2000.

Rohde, David. "It's Still a Fiber Game for Office Parks." *Network World Online*; August 31, 2000.

Rose, Dwane. "DWDM for Everyone." *Telephony*; June 5, 2000.

Saunders, Stephen. Village Unveils "Optical Packet Node." *Light Reading*; October 3, 2000.

Saunders, Stephen and Marguerite Reardon. "Organic Lasers: If it's Organic, it's Got to be Good." *Network World Online*; August 31, 2000.

Saunders, Stephen and Marguerite Reardon. "Making Optical Switching Crystal Clear." *Network World Online*; August 17, 2000.

Schagerlund, Olov. "Roadmap to Optical Networking." *Communications News*; November 1999.

Srikanth, Raj and Alex Barros. "Adding Fiber to Your Diet." *Upside*; July 2000.

Steinke, Steve. "Fundamentals of Optical Networking." *Network Magazine*; June 2000.

Steinke, Steve. Optical Networking and "Optical Networking." *Network Magazine*; June 2000.

Stewart, Alan. "The Great Lambda Wars." *America's Network*; May 1, 2000.

Swanson, Bret. "Circling the Fibersphere." *Gilder Technology Report*; July 2000, Volume V, Number 7.

Sweeney, Dan. "Mirrors and Smoke: The Optical Challenge." *America's Network*; June 1, 2000.

Taylor, Steve and Larry Hettick. "Core Migration: It's Mandatory." *Network World Online*; September 14, 2000.

Thomas, Gordon A., David A. Ackerman, Paul R. Prucnal, S. Lance Cooper. "Physics in the Whirlwind of Optical Communications." *Physics Today*; September 2000.

Wirbel, Loring. "Appian Adds Jewel to Optical Ring Protection." *EETimes*; August 28, 2000.

Wirbel, Loring. "Optical Pipes Get Thinner, Smarter with Demand." *EETimes*; August 30, 2000.

MISCELLANY

Notes from a series of discussions with Mitch Moore, senior member of the technical staff with Hill Associates, about the networking hierarchy of basic needs that progresses from hardware and software features to actual value as perceived by the beneficiary of the technology. Thanks, Mitch, for the insights and prodding.

COMMON INDUSTRY ACRONYMS

AAL	ATM Adaptation Layer
AARP	AppleTalk Address Resolution Protocol
ABM	Asynchronous Balanced Mode
ABR	Available Bit Rate
AC	Alternating Current
ACD	Automatic Call Distribution
ACELP	Algebraic Code Excited Linear Prediction
ACF	Advanced Communication Function
ACK	Acknowledgment
ACM	Address Complete Message
ACSE	Association Control Service Element
ACTLU	Activate Logical Unit
ACTPU	Activate Physical Unit
ADCCP	Advanced Data Communications Control Procedures
ADM	Add/Drop Multiplexer
ADPCM	Adaptive Differential Pulse Code Modulation
ADSL	Asymmetric Digital Subscriber Line

AFI	Authority and Format Identifier
AIN	Advanced Intelligent Network
AIS	Alarm Indication Signal
ALU	Arithmetic Logic Unit
AM	Administrative Module (Lucent 5ESS)
AM	Amplitude Modulation
AMI	Alternate Mark Inversion
AMP	Administrative Module Processor
AMPS	Advanced Mobile Phone System
ANI	Automatic Number Identification (SS7)
ANSI	American National Standards Institute
APD	Avalanche Photodiode
API	Application Programming Interface
APPC	Advanced Program-to-Program Communication
APPN	Advanced Peer-to-Peer Networking
APS	Automatic Protection Switching
ARE	All Routes Explorer (Source Route Bridging)
ARM	Asynchronous Response Mode
ARP	Address Resolution Protocol (IETF)
ARPA	Advanced Research Projects Agency
ARPANET	Advanced Research Projects Agency Network
ARQ	Automatic Repeat Request
ASCII	American Standard Code for Information Interchange
ASI	Alternate Space Inversion
ASIC	Application Specific Integrated Circuit
ASIC	Application-Specific Integrated Circuit
ASK	Amplitude Shift Keying
ASN	Abstract Syntax Notation

ASP	Application Service Provider
AT&T	American Telephone and Telegraph
ATDM	Asynchronous Time Division Multiplexing
ATM	Asynchronous Transfer Mode
ATM	Automatic Teller Machine
ATMF	ATM Forum
AWG	American Wire Gauge
B8ZS	Binary 8 Zero Substitution
BANCS	Bell Administrative Network Communications System
BBN	Bolt, Beranak, and Newman
BBS	Bulletin Board Service
Bc	Committed Burst Size
BCC	Blocked Calls Cleared
BCC	Block Check Character
BCD	Blocked Calls Delayed
BCDIC	Binary Coded Decimal Interchange Code
Be	Excess Burst Size
BECN	Backward Explicit Congestion Notification
BER	Bit Error Rate
BERT	Bit Error Rate Test
BGP	Border Gateway Protocol (IETF)
BIB	Backward Indicator Bit (SS7)
B-ICI	Broadband Intercarrier Interface
BIOS	Basic Input/Output System
BIP	Bit Interleaved Parity
B-ISDN	Broadband Integrated Services Digital Network
BISYNC	Binary Synchronous Communications Protocol
BITNET	Because It's Time Network
BITS	Building Integrated Timing Supply

BLSR	Bidirectional Line Switched Ring
BOC	Bell Operating Company
BPRZ	Bipolar Return to Zero
Bps	Bits per Second
BRI	Basic Rate Interface
BRITE	Basic Rate Interface Transmission Equipment
BSC	Binary Synchronous Communications
BSN	Backward Sequence Number (SS7)
BSRF	Bell System Reference Frequency
BTAM	Basic Telecommunications Access Method
BUS	Broadcast Unknown Server
C/R	Command/Response
CAD	Computer-Aided Design
CAE	Computer-Aided Engineering
CAM	Computer-Aided Manufacturing
CAP	Carrierless Amplitude/Phase Modulation
CAP	Competitive Access Provider
CARICOM	Caribbean Community and Common Market
CASE	Common Application Service Element
CASE	Computer-Aided Software Engineering
CAT	Computer-Aided Tomography
CATIA	Computer-Assisted Three-dimensional Interactive Application
CATV	Community Antenna Television
CBEMA	Computer and Business Equipment Manufacturers Association
CBR	Constant Bit Rate
CBT	Computer-Based Training
CC	Cluster Controller
CCIR	International Radio Consultative Committee

CCIS	Common Channel Interoffice Signaling
CCITT	International Telegraph and Telephone Consultative Committee
CCS	Common Channel Signaling
CCS	Hundred Call Seconds per Hour
CD	Collision Detection
CD	Compact Disc
CDC	Control Data Corporation
CDMA	Code Division Multiple Access
CDPD	Cellular Digital Packet Data
CD-ROM	Compact Disc-Read Only Memory
CDVT	Cell Delay Variation Tolerance
CEI	Comparably Efficient Interconnection
CEPT	Conference of European Postal and Telecommunications Administrations
CERN	European Council for Nuclear Research
CERT	Computer Emergency Response Team
CES	Circuit Emulation Service
CEV	Controlled Environmental Vault
CGI	Common Gateway Interface (Internet)
CHAP	Challenge Handshake Authentication Protocol
CICS	Customer Information Control System
CICS/VS	Customer Information Control System/Virtual Storage
CIDR	Classless Interdomain Routing (IETF)
CIF	Cells In Frames
CIR	Committed Information Rate
CISC	Complex Instruction Set Computer
CIX	Commercial Internet Exchange
CLASS	Custom Local Area Signaling Services

	(Bellcore)
CLEC	Competitive Local Exchange Carrier
CLLM	Consolidated Link Layer Management
CLNP	Connectionless Network Protocol
CLNS	Connectionless Network Service
CLP	Cell Loss Priority
CM	Communications Module (Lucent 5ESS)
CMIP	Common Management Information Protocol
CMISE	Common Management Information Service Element
CMOL	CMIP Over LLC
CMOS	Complementary Metal Oxide Semiconductor
CMOT	CMIP Over TCP/IP
CMP	Communications Module Processor
CNE	Certified NetWare Engineer
CNM	Customer Network Management
CNR	Carrier-to-Noise Ratio
CO	Central Office
CoCOM	Coordinating Committee on Export Controls
CODEC	Coder-Decoder
COMC	Communications Controller
CONS	Connection-Oriented Network Service
CORBA	Common Object Request Brokered Architecture
COS	Class of Service (APPN)
COS	Corporation for Open Systems
CPE	Customer Premises Equipment
CPU	Central Processing Unit
CRC	Cyclic Redundancy Check
CRT	Cathode Ray Tube

CRV	Call Reference Value
CS	Convergence Sublayer
CSA	Carrier Serving Area
CSMA	Carrier Sense Multiple Access
CSMA/CA	Carrier Sense Multiple Access with Collision Avoidance
CSMA/CD	Carrier Sense Multiple Access with Collision Detection
CSU	Channel Service Unit
CTI	Computer Telephony Integration
CTIA	Cellular Telecommunications Industry Association
CTS	Clear To Send
CU	Control Unit
CVSD	Continuously Variable Slope Delta modulation
CWDM	Coarse Wavelength Division Multiplexing
D/A	Digital-to-Analog
DA	Destination Address
DAC	Dual Attachment Concentrator (FDDI)
DACS	Digital Access and Cross-connect System
DARPA	Defense Advanced Research Projects Agency
DAS	Dual Attachment Station (FDDI)
DASD	Direct Access Storage Device
DB	Decibel
DBS	Direct Broadcast Satellite
DC	Direct Current
DCC	Data Communications Channel (SONET)
DCE	Data Circuit-terminating Equipment
DCN	Data Communications Network
DCS	Digital Cross-connect System

DCT	Discrete Cosine Transform
DDCMP	Digital Data Communications Management Protocol (DNA)
DDD	Direct Distance Dialing
DDP	Datagram Delivery Protocol
DDS	DATAPHONE Digital Service (Sometimes Digital Data Service)
DDS	Digital Data Service
DE	Discard Eligibility (LAPF)
DECT	Digital European Cordless Telephone
DES	Data Encryption Standard (NIST)
DID	Direct Inward Dialing
DIP	Dual Inline Package
DLC	Digital Loop Carrier
DLCI	Data Link Connection Identifier
DLE	Data Link Escape
DLSw	Data Link Switching
DM	Delta Modulation
DM	Disconnected Mode
DMA	Direct Memory Access (computers)
DMAC	Direct Memory Access Control
DME	Distributed Management Environment
DMS	Digital Multiplex Switch
DNA	Digital Network Architecture
DNIC	Data Network Identification Code (X.121)
DNIS	Dialed Number Identification Service
DNS	Domain Name System (IETF)
DOD	Direct Outward Dialing
DOD	Department of Defense
DOJ	Department of Justice

DOV	Data Over Voice
DPSK	Differential Phase Shift Keying
DQDB	Distributed Queue Dual Bus
DRAM	Dynamic Random Access Memory
DSAP	Destination Service Access Point
DSF	Dispersion-Shifted Fiber
DSI	Digital Speech Interpolation
DSL	Digital Subscriber Line
DSLAM	Digital Subscriber Line Access Multiplexer
DSP	Digital Signal Processing
DSR	Data Set Ready
DSS	Digital Satellite System
DSS	Digital Subscriber Signaling System
DSU	Data Service Unit
DTE	Data Terminal Equipment
DTMF	Dual Tone Multifrequency
DTR	Data Terminal Ready
DVRN	Dense Virtual Routed Networking (Crescent)
DWDM	Dense Wavelength Division Multiplexing
DXI	Data Exchange Interface
E/O	Electrical-to-Optical
EBCDIC	Extended Binary Coded Decimal Interchange Code
ECMA	European Computer Manufacturer Association
ECN	Explicit Congestion Notification
ECSA	Exchange Carriers Standards Association
EDFA	Erbium-Doped Fiber Amplifier
EDI	Electronic Data Interchange
EDIBANX	EDI Bank Alliance Network Exchange
EDIFACT	Electronic Data Interchange For

	Administration, Commerce, and Trade (ANSI)
EFCI	Explicit Forward Congestion Indicator
EFTA	European Free Trade Association
EGP	Exterior Gateway Protocol (IETF)
EIA	Electronics Industry Association
EIGRP	Enhanced Interior Gateway Routing Protocol
EIR	Excess Information Rate
EMBARC	Electronic Mail Broadcast to a Roaming Computer
EMI	Electromagnetic Interference
EMS	Element Management System
EN	End Node
ENIAC	Electronic Numerical Integrator And Computer
EO	End Office
EOC	Embedded Operations Channel (SONET)
EOT	End Of Transmission (BISYNC)
EPROM	Erasable Programmable Read Only Memory
ESCON	Enterprise System Connection (IBM)
ESF	Extended Superframe Format
ESP	Enhanced Service Provider
ESS	Electronic Switching System
ETSI	European Telecommunications Standards Institute
ETX	End of Text (BISYNC)
EWOS	European Workshop for Open Systems
FACTR	Fujitsu Access and Transport System
FAQ	Frequently Asked Questions
FAT	File Allocation Table
FCS	Frame Check Sequence

FDD	Frequency Division Duplex
FDDI	Fiber Distributed Data Interface
FDM	Frequency Division Multiplexing
FDMA	Frequency Division Multiple Access
FDX	Full-Duplex
FEBE	Far End Block Error (SONET)
FEC	Forward Error Correction
FEC	Forward Equivalence Class
FECN	Forward Explicit Congestion Notification
FEP	Front-End Processor
FERF	Far End Receive Failure (SONET)
FET	Field Effect Transistor
FHSS	Frequency Hopping Spread Spectrum
FIB	Forward Indicator Bit (SS7)
FIFO	First In First Out
FITL	Fiber In The Loop
FLAG	Fiber Link Across the Globe
FM	Frequency Modulation
FPGA	Field Programmable Gate Array
FR	Frame Relay
FRAD	Frame Relay Access Device
FRBS	Frame Relay Bearer Service
FSK	Frequency Shift Keying
FSN	Forward Sequence Number (SS7)
FTAM	File Transfer, Access, and Management
FTP	File Transfer Protocol (IETF)
FTTC	Fiber to the Curb
FTTH	Fiber to the Home
FUNI	Frame User-to-Network Interface
FWM	Four Wave Mixing

GATT	General Agreement on Tariffs and Trade
GbE	Gigabit Ethernet
Gbps	Gigabits per Second (Billion bits per second)
GDMO	Guidelines for the Development of Managed Objects
GEOS	Geosynchronous Earth Orbit Satellites
GFC	Generic Flow Control (ATM)
GFI	General Format Identifier (X.25)
GOSIP	Government Open Systems Interconnection Profile
GPS	Global Positioning System
GRIN	Graded Index (fiber)
GSM	Global System for Mobile Communications
GUI	Graphical User Interface
HDB3	High Density, Bipolar 3 (E-Carrier)
HDLC	High-level Data Link Control
HDSL	High-bit-rate Digital Subscriber Line
HDTV	High Definition Television
HDX	Half-Duplex
HEC	Header Error Control (ATM)
HFC	Hybrid Fiber/Coax
HFS	Hierarchical File Storage
HLR	Home Location Register
HSSI	High-Speed Serial Interface (ANSI)
HTML	Hypertext Markup Language
HTTP	Hypertext Transfer Protocol (IETF)
HTU	HDSL Transmission Unit
I	Intrapictures
IAB	Internet Architecture Board (formerly Internet Activities Board)
IACS	Integrated Access and Cross-connect System

IAD	Integrated Access Device
IAM	Initial Address Message (SS7)
IANA	Internet Address Naming Authority
ICMP	Internet Control Message Protocol (IETF)
IDP	Internet Datagram Protocol
IEC	Interexchange Carrier (also IXC)
IEC	International Electrotechnical Commission
IEEE	Institute of Electrical and Electronics Engineers
IETF	Internet Engineering Task Force
IFRB	International Frequency Registration Board
IGP	Interior Gateway Protocol (IETF)
IGRP	Interior Gateway Routing Protocol
ILEC	Incumbent Local ExchangeCarrier
IML	Initial Microcode Load
IMP	Interface Message Processor (ARPANET)
IMS	Information Management System
InARP	Inverse Address Resolution Protocol (IETF)
InATMARP	Inverse ATMARP
INMARSAT	International Maritime Satellite Organization
INP	Internet Nodal Processor
InterNIC	Internet Network Information Center
IP	Internet Protocol (IETF)
IPX	Internetwork Packet Exchange (NetWare)
ISDN	Integrated Services Digital Network
ISO	International Organization for Standardization
ISOC	Internet Society
ISP	Internet Service Provider
ISUP	ISDN User Part (SS7)
IT	Information Technology

ITU	International Telecommunication Union
ITU-R	International Telecommunication Union-Radio Communication Sector
IVD	Inside Vapor Deposition
IVR	Interactive Voice Response
IXC	Interexchange Carrier
JEPI	Joint Electronic Paynets Initiative
JES	Job Entry System
JIT	Just in Time
JPEG	Joint Photographic Experts Group
KB	Kilobytes
Kbps	Kilobits per Second (Thousand Bits per Second)
KLTN	Potassium Lithium Tantalate Niobate
LAN	Local Area Network
LANE	LAN Emulation
LAP	Link Access Procedure (X.25)
LAPB	Link Access Procedure Balanced (X.25)
LAPD	Link Access Procedure for the D-Channel
LAPF	Link Access Procedure to Frame Mode Bearer Services
LAPF-Core	Core Aspects of the Link Access Procedure to Frame Mode Bearer Services
LAPM	Link Access Procedure for Modems
LAPX	Link Access Procedure half-duplex
LASER	Light Amplification by the Stimulated Emission of Radiation
LATA	Local Access and Transport Area
LCD	Liquid Crystal Display
LCGN	Logical Channel Group Number
LCM	Line Concentrator Module

LCN	Local Communications Network
LD	Laser Diode
LDAP	Lightweight Directory Access Protocol (X.500)
LEAF®	Large Effective Area Fiber® (Corning product)
LEC	Local Exchange Carrier
LED	Light Emitting Diode
LENS	Lightwave Efficient Network Solution (Centerpoint)
LEOS	Low Earth Orbit Satellites
LI	Length Indicator
LIDB	Line Information Database
LIFO	Last In First Out
LIS	Logical IP Subnet
LLC	Logical Link Control
LMDS	Local Multipoint Distribution System
LMI	Local Management Interface
LMOS	Loop Maintenance Operations System
LORAN	Long-range Radio Navigation
LPC	Linear Predictive Coding
LPP	Lightweight Presentation Protocol
LRC	Longitudinal Redundancy Check (BISYNC)
LS	Link State
LSI	Large Scale Integration
LSP	Label Switched Path
LU	Line Unit
LU	Logical Unit (SNA)
MAC	Media Access Control
MAN	Metropolitan Area Network
MAP	Manufacturing Automation Protocol
MAU	Medium Attachment Unit (Ethernet)

MAU	Multistation Access Unit (Token Ring)
MB	Megabytes
MBA™	Metro Business Access™ (Ocular)
Mbps	Megabits per Second (Million bits per second)
MD	Message Digest (MD2, MD4, MD5) (IETF)
MDF	Main Distribution Frame
MEMS	Micro Electrical Mechanical System
MF	Multifrequency
MFJ	Modified Final Judgment
MHS	Message Handling System (X.400)
MIB	Management Information Base
MIC	Medium Interface Connector (FDDI)
MIME	Multipurpose Internet Mail Extensions (IETF)
MIPS	Millions of Instructions Per Second
MIS	Management Information Systems
MITI	Ministry of International Trade and Industry (Japan)
ML-PPP	Multilink Point-to-Point Protocol
MMDS	Multichannel, Multipoint Distribution System
MMF	Multimode Fiber
MNP	Microcom Networking Protocol
MP	Multilink PPP
MPEG	Moving Picture Experts Group
MPLS	Multiprotocol Label Switching
MPOA	Multiprotocol Over ATM
MRI	Magnetic Resonance Imaging
MSB	Most Significant Bit
MSC	Mobile Switching Center
MSO	Mobile Switching Office
MSVC	Meta-Signaling Virtual Channel

MTA	Major Trading Area
MTBF	Mean Time Between Failure
MTP	Message Transfer Part (SS7)
MTTR	Mean Time To Repair
MTU	Maximum Transmission Unit
MVS	Multiple Virtual Storage
NAFTA	North American Free Trade Agreement
NAK	Negative Acknowledgment (BISYNC, DDCMP)
NAP	Network Access Point (Internet)
NARUC	National Association of Regulatory Utility Commissioners
NASA	National Aeronautics and Space Administration
NASDAQ	National Association of Securities Dealers Automated Quotations
NATA	North American Telecommunications Association
NATO	North Atlantic Treaty Organization
NAU	Network Accessible Unit
NCP	Network Control Program
NCSA	National Center for Supercomputer Applications
NCTA	National Cable Television Association
NDIS	Network Driver Interface Specifications
NDSF	Non-Dispersion-Shifted Fiber
NetBEUI	NetBIOS Extended User Interface
NetBIOS	Network Basic Input/Output System
NFS	Network File System (Sun)
NIC	Network Interface Card
NII	National Information Infrastructure
NIST	National Institute of Standards and Technology

(formerly NBS)

NIU	Network Interface Unit
NLPID	Network Layer Protocol Identifier
NLSP	NetWare Link Services Protocol
NM	Network Module
Nm	Nanometer
NMC	Network Management Center
NMS	Network Management System
NMT	Nordic Mobile Telephone
NMVT	Network Management Vector Transport protocol
NNI	Network Node Interface
NNI	Network-to-Network Interface
NOC	Network Operations Center
NOCC	Network Operations Control Center
NOS	Network Operating System
NPA	Numbering Plan Area
NREN	National Research and Education Network
NRZ	Non-Return to Zero
NRZI	Non-Return to Zero Inverted
NSA	National Security Agency
NSAP	Network Service Access Point
NSAPA	Network Service Access Point Address
NSF	National Science Foundation
NTSC	National Television Systems Committee
NTT	Nippon Telephone and Telegraph
NVOD	Near Video On Demand
NZDSF	Non-Zero Dispersion-Shifted Fiber
OADM	Optical Add-Drop Multiplexer
OAM	Operations, Administration, and Maintenance

OAM&P	Operations, Administration, Maintenance, and Provisioning
OAN	Optical Area Network
OC	Optical Carrier
OEM	Original Equipment Manufacturer
O-E-O	Optical-Electrical-Optical
OLS	Optical Line System (Lucent)
OMAP	Operations, Maintenance, and Administration Part (SS7)
ONA	Open Network Architecture
ONU	Optical Network Unit
OOF	Out Of Frame
OS	Operating System
OSF	Open Software Foundation
OSI	Open Systems Interconnection (ISO, ITU-T)
OSI RM	Open Systems Interconnection Reference Model
OSPF	Open Shortest Path First (IETF)
OSS	Operation Support Systems
OTDM	Optical Time Division Multiplexing
OTDR	Optical Time-Domain Reflectometer
OUI	Organizationally Unique Identifier (SNAP)
OVD	Outside Vapor Deposition
P/F	Poll/Final (HDLC)
PAD	Packet Assembler/Disassembler (X.25)
PAL	Phase Alternate Line
PAM	Pulse Amplitude Modulation
PANS	Pretty Amazing New Stuff
PBX	Private Branch Exchange
PCI	Pulse Code Modulation

PCMCIA	Personal Computer Memory Card International Association
PCN	Personal Communications Network
PCS	Personal Communications Services
PDA	Personal Digital Assistant
PDU	Protocol Data Unit
PIN	Positive-Intrinsic-Negative
PING	Packet Internet Groper (TCP/IP)
PLCP	Physical Layer Convergence Protocol
PLP	Packet Layer Protocol (X.25)
PM	Phase Modulation
PMD	Physical Medium Dependent (FDDI)
PNNI	Private Network Node Interface (ATM)
PON	Passive Optical Networking
POP	Point Of Presence
POSIT	Profiles for Open Systems Interworking Technologies
POSIX	Portable Operating System Interface for UNIX
POTS	Plain Old Telephone Service
PPP	Point-to-Point Protocol (IETF)
PRC	Primary Reference Clock
PRI	Primary Rate Interface
PROFS	Professional Office System
PROM	Programmable Read Only Memory
PSDN	Packet Switched Data Network
PSK	Phase Shift Keying
PSPDN	Packet Switched Public Data Network
PSTN	Public Switched Telephone Network
PTI	Payload Type Identifier (ATM)
PTT	Post, Telephone, and Telegraph

PU	Physical Unit (SNA)
PUC	Public Utility Commission
PVC	Permanent Virtual Circuit
QAM	Quadrature Amplitude Modulation
Q-bit	Qualified data bit (X.25)
QLLC	Qualified Logical Link Control (SNA)
QoS	Quality of Service
QPSK	Quadrature Phase Shift Keying
QPSX	Queued Packet Synchronous Exchange
R&D	Research & Development
RADSL	Rate Adaptive Digital Subscriber Line
RAID	Redundant Array of Inexpensive Disks
RAM	Random Access Memory
RARP	Reverse Address Resolution Protocol (IETF)
RAS	Remote Access Server
RBOC	Regional Bell Operating Company
RF	Radio Frequency
RFC	Request For Comments (IETF)
RFH	Remote Frame Handler (ISDN)
RFI	Radio Frequency Interference
RFP	Request For Proposal
RHC	Regional Holding Company
RHK	Ryan, Hankin and Kent (Consultancy)
RIP	Routing Information Protocol (IETF)
RISC	Reduced Instruction Set Computer
RJE	Remote Job Entry
RNR	Receive Not Ready (HDLC)
ROM	Read-Only Memory
ROSE	Remote Operation Service Element
RPC	Remote Procedure Call

RR	Receive Ready (HDLC)
RTS	Request To Send (EIA-232-E)
S/DMS	SONET/Digital Multiplex System
S/N	Signal-to-Noise Ratio
SAA	Systems Application Architecture (IBM)
SAAL	Signaling ATM Adaptation Layer (ATM)
SABM	Set Asynchronous Balanced Mode (HDLC)
SABME	Set Asynchronous Balanced Mode Extended (HDLC)
SAC	Single Attachment Concentrator (FDDI)
SAN	Storage Area Network
SAP	Service Access Point (generic)
SAPI	Service Access Point Identifier (LAPD)
SAR	Segmentation And Reassembly (ATM)
SAS	Single Attachment Station (FDDI)
SASE	Specific Applications Service Element (subset of CASE, Application Layer)
SATAN	System Administrator Tool for Analyzing Networks
SBS	Stimulated Brillouin Scattering
SCCP	Signaling Connection Control Point (SS7)
SCP	Service Control Point (SS7)
SCREAM™	Scalable Control of a Rearrangeable Extensible Array of Mirrors (Calient)
SCSI	Small Computer Systems Interface
SCTE	Serial Clock Transmit External (EIA-232-E)
SDH	Synchronous Digital Hierarchy (ITU-T)
SDLC	Synchronous Data Link Control (IBM)
SDS	Scientific Data Systems
SECAM	Sequential Color with Memory

SF	Superframe Format (T-1)
SGML	Standard Generalized Markup Language
SGMP	Simple Gateway Management Protocol (IETF)
S-HTTP	Secure HTTP (IETF)
SIF	Signaling Information Field
SIG	Special Interest Group
SIO	Service Information Octet
SIR	Sustained Information Rate (SMDS)
SLA	Service Level Agreement
SLIP	Serial Line Interface Protocol (IETF)
SM	Switching Module
SMAP	System Management Application Part
SMDS	Switched Multimegabit Data Service
SMF	Single Mode Fiber
SMP	Simple Management Protocol
SMP	Switching Module Processor
SMR	Specialized Mobile Radio
SMS	Standard Management System (SS7)
SMTP	Simple Mail Transfer Protocol (IETF)
SNA	Systems Network Architecture (IBM)
SNAP	Subnetwork Access Protocol
SNI	Subscriber Network Interface (SMDS)
SNMP	Simple Network Management Protocol (IETF)
SNP	Sequence Number Protection
SONET	Synchronous Optical Network
SPAG	Standards Promotion and Application Group
SPARC	Scalable Performance Architecture
SPE	Synchronous Payload Envelope (SONET)
SPID	Service Profile Identifier (ISDN)
SPM	Self Phase Modulation

SPOC	Single Point Of Contact
SPX	Sequenced Packet Exchange (NetWare)
SQL	Structured Query Language
SRB	Source Route Bridging
SRS	Stimulated Raman Scattering
SRT	Source Routing Transparent
SS7	Signaling System 7
SSL	Secure Socket Layer (IETF)
SSP	Service Switching Point (SS7)
SST	Spread Spectrum Transmission
STDM	Statistical Time Division Multiplexing
STM	Synchronous Transfer Mode
STM	Synchronous Transport Module (SDH)
STP	Signal Transfer Point (SS7)
STS	Synchronous Transport Signal (SONET)
STX	Start of Text (BISYNC)
SVC	Signaling Virtual Channel (ATM)
SVC	Switched Virtual Circuit
SXS	Step-by-Step Switching
SYN	Synchronization
SYNTRAN	Synchronous Transmission
TA	Terminal Adapter (ISDN)
TAG	Technical Advisory Group
TASI	Time Assigned Speech Interpolation
TAXI	Transparent Asynchronous Transmitter/Receiver Interface (Physical Layer)
TCAP	Transaction Capabilities Application Part (SS7)
TCM	Time Compression Multiplexing
TCM	Trellis Coding Modulation
TCP	Transmission Control Protocol (IETF)

TDD	Time Division Duplexing
TDM	Time Division Multiplexing
TDM	Time Division Multiplexing
TDMA	Time Division Multiple Access
TDR	Time Domain Reflectometer
TE1	Terminal Equipment type 1 (ISDN capable)
TE2	Terminal Equipment type 2 (non-ISDN capable)
TEI	Terminal Endpoint Identifier (LAPD)
TELRIC	Total Element Long-Run Incremental Cost
TIA	Telecommunications Industry Association
TIRKS	Trunk Integrated Record Keeping System
TL1	Transaction Language 1
TM	Terminal Multiplexer
TMN	Telecommunications Management Network
TMS	Time-Multiplexed Switch
TOH	Transport Overhead (SONET)
TOP	Technical and Office Protocol
TOS	Type Of Service (IP)
TP	Twisted Pair
TR	Token Ring
TRA	Traffic Routing Administration
TSI	Time Slot Interchange
TSLRIC	Total Service Long-Run Incremental Cost
TSO	Terminating Screening Office
TSO	Time-Sharing Option (IBM)
TSR	Terminate and Stay Resident
TSS	Telecommunication Standardization Sector (ITU-T)
TST	Time-Space-Time Switching

TSTS	Time-Space-Time-Space Switching
TTL	Time To Live
TUP	Telephone User Part (SS7)
UA	Unnumbered Acknowledgment (HDLC)
UART	Universal Asynchronous Receiver Transmitter
UBR	Unspecified Bit Rate (ATM)
UDI	Unrestricted Digital Information (ISDN)
UDP	User Datagram Protocol (IETF)
UHF	Ultra High Frequency
UI	Unnumbered Information (HDLC)
UNI	User-to-Network Interface (ATM, FR)
UNIT™	Unified Network Interface Technology™ (Ocular)
UNMA	Unified Network Management Architecture
UPS	Uninterruptable Power Supply
UPSR	Unidirectional Path Switched Ring
UPT	Universal Personal Telecommunications
URL	Uniform Resource Locator
USART	Universal Synchronous Asynchronous Receiver Transmitter
UTC	Coordinated Universal Time
UTP	Unshielded Twisted Pair (Physical Layer)
UUCP	UNIX-UNIX Copy
VAN	Value-Added Network
VAX	Virtual Address Extension (DEC)
vBNS	Very High Speed Backbone Network Service
VBR	Variable Bit Rate (ATM)
VBR-NRT	Variable Bit Rate-Non-Real-Time (ATM)
VBR-RT	Variable Bit Rate-Real-Time (ATM)
VC	Virtual Channel (ATM)

VC	Virtual Circuit (PSN)
VCC	Virtual Channel Connection (ATM)
VCI	Virtual Channel Identifier (ATM)
VCI	Virtual Channel Identifier (ATM)
VCSEL	Vertical Cavity Surface Emitting Laser
VDSL	Very High-Speed Digital Subscriber Line
VDSL	Very High bit rate Digital Subscriber Line
VERONICA	Very Easy Rodent-Oriented Netwide Index to Computerized Archives (Internet)
VGA	Variable Graphics Array
VHF	Very High Frequency
VHS	Video Home System
VINES	Virtual Networking System (Banyan)
VIP	VINES Internet Protocol
VLF	Very Low Frequency
VLR	Visitor Location Register (Wireless/GSM)
VLSI	Very Large Scale Integration
VM	Virtual Machine (IBM)
VM	Virtual Memory
VMS	Virtual Memory System (DEC)
VOD	Video-On-Demand
VP	Virtual Path
VPC	Virtual Path Connection
VPI	Virtual Path Identifier
VPN	Virtual Private Network
VPN	Virtual Private Network
VR	Virtual Reality
VSAT	Very Small Aperture Terminal
VSB	Vestigial Sideband
VSELP	Vector-Sum Excited Linear Prediction

VT	Virtual Tributary
VTAM	Virtual Telecommunications Access Method (SNA)
VTOA	Voice and Telephony Over ATM
VTP	Virtual Terminal Protocol (ISO)
WACK	Wait Acknowledgment (BISYNC)
WACS	Wireless Access Communications System
WAIS	Wide Area Information Server (IETF)
WAN	Wide Area Network
WARC	World Administrative Radio Conference
WATS	Wide Area Telecommunications Service
WDM	Wavelength Division Multiplexing
WIN	Wireless In-building Network
WTO	World Trade Organization
WWW	World Wide Web (IETF)
WYSIWYG	What You See Is What You Get
xDSL	x-Type Digital Subscriber Line
XID	Exchange Identification (HDLC)
XNS	Xerox Network Systems
XPM	Cross Phase Modulation
ZBTSI	Zero Byte Time Slot Interchange
ZCS	Zero Code Suppression

INDEX

ABOUT THE AUTHOR

Steven Shepard is a professional writer and educator, specializing in international telecommunications. He teaches technical courses in corporations throughout the world. Formerly with Hill Associates, he is the author of *Telecommunications Convergence*, also from McGraw-Hill. He is based in Williston, Vermont.